Statistical Pattern Classification using Contextual Information

ELECTRONIC & ELECTRICAL ENGINEERING RESEARCH STUDIES

PATTERN RECOGNITION & IMAGE PROCESSING SERIES

Series Editor: **Dr. Josef Kittler,** *Balliol College, Oxford*

1: Statistical Pattern Classification using Contextual Information

Professor K-S Fu *and* **Dr. T.S. Yu**

Statistical Pattern Classification using Contextual Information

Professor K-S Fu
Goss Professor of Engineering,
School of Electrical Engineering,
Purdue University, West Lafayette,
Indiana, U.S.A.

and

Dr. T.S. Yu
Control Data Corporation,
Minneapolis, Minnesota, U.S.A.

RESEARCH STUDIES PRESS
A Division of John Wiley & Sons Ltd
Chichester · New York · Brisbane · Toronto

RESEARCH STUDIES PRESS

Editorial Office:
8 Willian Way, Letchworth, Herts SG6 2HG, England

British Library Cataloguing in Publication Data:

Statistical pattern classification using contextual information.
- (Electronic and electrical engineering research studies: pattern recognition and image processing series: vol. 1).
1. Pattern recognition systems
I. Fu, King Sun
II. Yu, T.S.
III. Series
621.3819'598 TK7882.P3 80-40949

ISBN 0 471 27859 9

Printed in the United States of America

TABLE OF CONTENTS

PREFACE

In the past decade, there have been numerous attempts to develop systems for automatic interpretation of digital image data. None of the systems developed for this purpose have made extensive use of context information. Since even manual interpretation of isolated point or area targets is difficult without the use of context, a machine which does not use context has a fundamental limitation.

First, compound decision rule is adopted as the preliminary approach to introduce contextual information into the classification process. Based on initial success in using context information, the investigation has been extended and is going beyond the simple (no context) classification to improve the decision procedure in areas where the need is clear and the possibilities are promising. The net outcome of this investigation has been a conclusive demonstration of the feasibility of applying a spatial stochastic model to analyze digital Satellite data. The effectiveness of this model is demonstrated by the reasonable increase in computation time and the significantly improved classification results.

In the course of the investigation, the first task is to develop a stationary stochastic process on a two-dimensional plane. This process is then used as a model such that correlations between any pair of image cells can be extracted. The model is characterized by a spatial correlation parameter. A very flexible coding technique is presented by which the spatial correlation parameter can be estimated. Some modifications and extensions of the model are specifically developed or substantially refined during this investigation to cover more general situations.

From the coding pattern used for estimating the spatial correlation parameter, a recursive contextual classification algorithm is that the developed model can be fitted by the image data more appropriately.

Finally, extensive experimental results with remotely sensed multispectral scanner data using the developed model for contextual classification are reported. Both single-stage and recursive contextual classification procedures are tested on real data. The classification results do show the effectiveness and efficiency of the developed contextual classifier.

CHAPTER 1

INTRODUCTION

1.1 Pattern Recognition and Remote Sensing

There has been a considerable interest in machine processing of remotely sensed multispectral scanner (MSS) data in the past decade. Remote sensing is the acquisition and interpretation of spectral measurements made at a distant location to obtain information about the earth's surface. Pattern recognition techniques (12),(13,(78) have been extensively applied to interpret the data.

The physical basis for remote sensing is the distinctive character of electromagnetic radiance from natural and man-made scenes (84). Recent studies indicate that remote sensing techniques can be used in the visible, infrared, and microwave region of the electromagnetic spectrum to collect measurements of reflectance and emittance of plants, soils, water, and other materials. With a minimum amount of ground truth information, remote sensing data will permit identification (classification) and area measurements of crop, assessment of crop condition and yield potential, land use mapping, and surveys of water and geological resources.

The supply of remotely sensed data greatly increased with the launch of LANDSAT-1 (formerly called the Earth Resources Technology Satellite or ERTS-1) in 1972. From an orbit 912 Km above the earth, this satellite can complete a full observation of the earth every 18 days. Its multispectral imagery is collected in four visible and infrared wavelength bands over 185 Km wide passes over the earth. This newest data source has opened a whole new dimension to our capability to obtain information about earth resources.

Current procedures for digitally processing and analyzing data from LANDSAT, SKYLAB, or other multispectral scanner systems involve four primary areas of activity. These include data reformatting and preprocessing, analysis and classification of data, information display and tabulation, and, finally, evaluation of the results. Data reformatting and preprocessing procedures do not involve any actual analysis of the data, but simply allow subsequent data analysis activities to be carried out in a much more effective manner.

The classification of MSS data involves a series of pattern recognition steps to enable the computer to identify various crop types or earth surface features of interest. The key element in such computer aided analysis techniques involves a man/machine interaction, whereby the man "trains" the computer recognize particular combinations of numbers that represent reflectance measurements in each of several wavelength bands, for particular cover types of interest. This training processing is carried out utilizing scanner data collected over a limited geographic area. Then, after a good set of training statistics have been computed, the computer is programmed to classify the reflectance values for each resolution element in the entire data set. In this way, the computer can map and tabulate cover types over a large geographic area at a much faster rate than would be possible for a man using standard image interpretation techniques.

The first consideration in computer-aided analysis of MSS data involves the definition of the categories or classes of material that the computer should be trained to recognize. Basically, there are two conditions which must be met by each class involved in an analysis of MSS data using these computer-aided analysis techniques:

- The class must be spectrally separable from all other classes,
- The class must be of interest to the user or have informational value.

Quite often one finds in working with MSS data that the classes of interest to the user can not be spectrally separated. The need for a class to be both separable and to have informational values leads to two quite different basic approaches in training the computer.

The first approach is referred to as the "supervised technique," and involves use of a system of X-Y coordinates to designate to the computer the locations of known earth surface features that have informational value. However, experience has shown that for some earth cover types such as wildland areas, where the spectral responses are not homogeneous, this supervised technique often does not enable adequate accuracy to be achieved.

A second approach to training the computer involves the "clustering" technique (referred to as the "non-supervised" technique). in this approach the analyst simply designates the number of spectrally distinct classes into which the data to be classified should be divided. The computer is programmed to classify the data into the designated number of spectral classes and then prints out a map indicating which resolution elements in the data belong to which spectral classes. The analyst then relates this classification output map to known surface observation data, and determines which materials are represented by each of the different spectral classes indicated on the map. Figure 1.1 shows a flow chart of the clustering algorithm (9).

After the training statistics are defined, the maximum likelihood (ML) decision rule is used to classify the entire data set. Test fields can be obtained from the data with reference to "ground truth" information which is usually represented by two dimensional maps (e.g. USGS Topographic Maps) and aerial photographs. Classification accuracy can then be assessed using those test fields. This is a relatively simple pattern recognition algorithm, and has been used extensively at LARS* and elsewhere for analysis of remotely sensored data. However, much

---

* LARS (Laboratory for Applications of Remote Sensing, Purdue University).

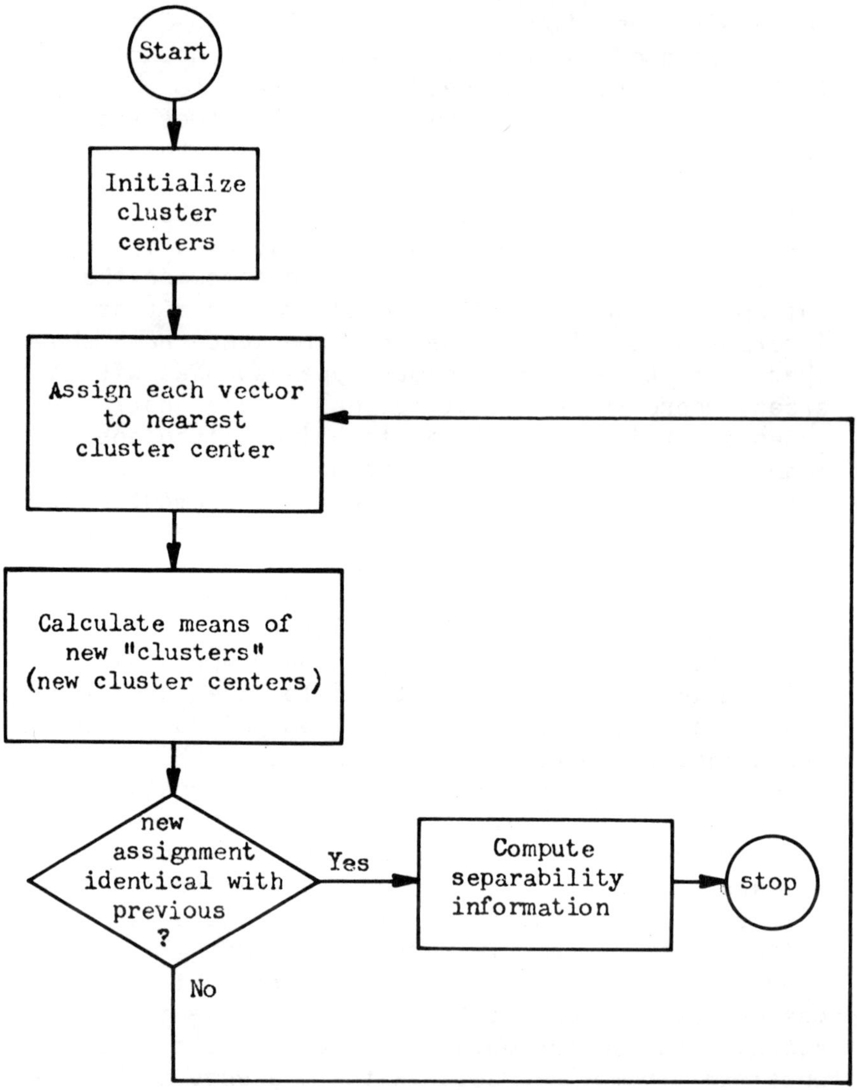

Figure 1.1 Flow chart of a clustering algorithm.

research work in the area of clustering and classification techniques (84-86) has been pursued to develop more efficient and accurate pattern recognition algorithms.

The objective of this study is to develop a class of contextual classifiers which make use of contextual information in the decision making process. Since even an amateur photo interpreter would concur that the contextual information component is substantial, a recognition system which does not use context in making decisions definitely has a fundamental handicap. The practical application of the proposed classifier is investigated for pattern recognition problems encountered in multispectral remote sensing.

The conventional approach to multivariate and multiclass classification would be to perform tests on the unknown sample against all classes using a particular feature subset and then assign the unknown to one of these classes. the discriminant functions used in performing tests on the unknown do not contain any contextual information for either the maximum likelihood or the Bayes (minimum risk) decision rule. However, in the proposed classifier, the contextual information is incorporated into the discriminant function such that this information can be used in a constructive way.

In almost all cases, the identities (true classes) of the neighboring patterns are also unknown to us. The question which immediately arises is how do we use information of which we are not sure. One way to solve this problem is to make an estimate of the true identities of the neighboring unknown patterns and then use these estimated true classes to provide contextual information. In this case, the true class transition probability context is employed. For example, automobiles occur more often on highways than in water, and boats are more often in water than on highways. Then, even if their spectral appearance is the same, boats can be discriminated from automobiles if their context is considered. The data immediately surrounding an image point (pixel) is intimately associated with it and can be used to throw light upon its true nature once we know the (estimated) true nature (class) of the surrounding pixels. We shall see in Chapter 2 how compound decision rules can provide class transition probability context.

The second type of contextual information is the correlation between neighboring patterns in image data. Every pixel is considered to be a random variable taking values according to one of known class conditional probability density functions. Therefore the random reflectance values of the surrounding pixels are definitely correlated with the center pixel being considered. It is a very challenging task to develop a probability model to extract this (spatial) correlation information in the two dimensional plane. We shall see in Chapter 3 how to develop and apply the spatial correlation type contextual information in a classification scheme.

Our reason for pursuing this investigation of the contextual decision rules is based on the advantages these decision rules may have. Three major advantages have been found, namely, high classification accuracy, reasonable computation time and more meaningful interpretation of results.

The main obstacle to implementing the contextual classifier is the difficulty in developing practical decision rules. To develop practical contextual rules and to demonstrate theoretically or experimentally the advantages of these rules thus become the major goals in this investigation.

1.2 Review of Related Work

Most of the literature of pattern recognition deals with simple (no context) Bayes classifiers. For a broad understanding of various pattern recognition techniques, readers may refer to the books by Duda and Hart (75), Chen (76), Fukunaga (82), Meisel (81), and Mendel and Fu (74). The survey papers by Fu (13), Fu and Rosenfeld (78), Ho and Agrawala (79), Kanal (80), and Nagy (77) are also helpful. For multispectral pattern recognition problems, a very complete survey is given by Nagy (77).

Most of the contextual classification work has been done in the area of character recognition (2,14,15) using compound or sequential compound decision theory. The data encountered in the character recognition problem usually come in the form of a one dimensional string. Raviv (2) uses this fact to make a Markov assumption and arrives at a recursive sequential compound decision rule. Direct application of Raviv's contextual character recognition algorithm to two dimensional remote sensing MSS data will fail since any Markov definition in the one dimensional case will not fit the two dimensional data. In fact a new body of theory on two or more dimensional stochastic Markov processes will have to be developed to handle imagery data. Chapter 3 presents this problem and its solution in a more detailed manner.

Many attempts have been made to improve the conventional maximum likelihood or Bayes classifier for classification of LANDSAT MSS data. However, little work has been done to employ statistical contextual information for classification of MSS data.

Wu (87) presented a multistage classification strategy called the decision tree classifier. The decision tree classifier is characterized by the fact that an unknown sample is classified into a class using one or several decision function in a successive manner. This classification strategy can be depicted by a tree diagram (Figure 1.2). To achieve the best possible performance with a classifier of this type, the design of the decision stages is of utmost importance. The choice of tree structure and the choice of appropriate feature subsets used at every "node" will be reflected in the performance (classification accuracy) and efficiency (computation time used). Wu used a Gaussian maximum likelihood rule at each stage of the tree.

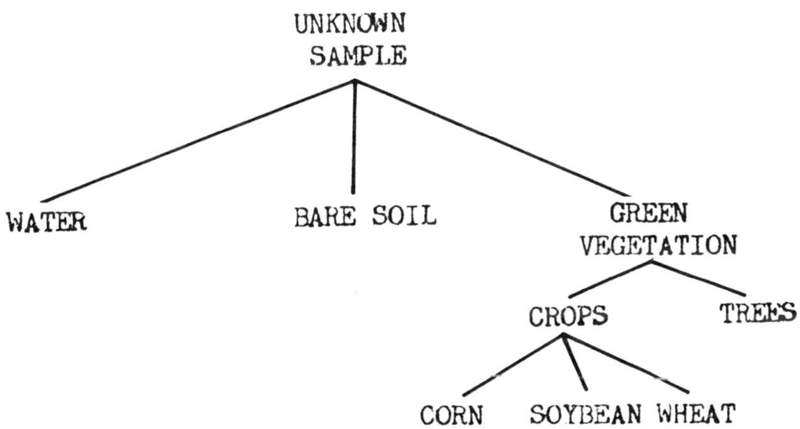

Figure 1.2 A simple decision tree.

If we restrict the number of features (spectral bands) used in each stage to one, we can manually design the tree structure by estimating suitable decision boundaries from the "coincident spectral plot" such that all classes are separated in a number of decision steps (86). The coincident spectral plot is a graph in which the means and variances for all the classes are plotted for each feature.

If the difficulty of discriminating the classes requires use of a combination of several features, the manual design approach based on the spectral plot is severely limited. In general, a more analytical multivariate design procedure is desirable when the complexity of the problem is significant.

For more complex problems, a heuristic search technique described as "guided search with forward pruning" (86,87) is used to design the tree. This method uses an evaluation function to direct a search through the possible decision tree structures such that, at each stage, the node configuration selected is the one with the highest evaluation function (112,113).

The evaluation function used in the search is a weighted measure of classifier efficiency and accuracy; i.e., the evaluation function of a given node d_i is of the form

$$E(d_i) = -T(d_i) - K \cdot \varepsilon\,(d_i) + \sum_{j=1}^{c_i} P_{i+j}\; E(d_{i+j}) \quad (1.1)$$

Where $T(d_i)$ is the computation time and $\varepsilon(d_i)$ is the classification error associated with the node; $E(d_{i+j})$ is the evaluation function associated with the jth descendant node of d_i; node of d_i and C_i descendant nodes; and P_{i+j} is the probability that the jth descendant node will be reached from d_i. The constant K is specified by the user to express the relative importance of classifier speed versus accuracy (112).

When the maximum number of features used per node

is four, experimental results showed that the decision tree classification time is about one-third of the single-stage classification time (86). However, when the number of features used per node is two, the decision tree classification time becomes two-thirds of that of the conventional classifier. LANDSAT data are collected in four wavelength bands. However only two bands have informational value because the two bands in each of the visible and infrared region provide essentially the same information. Therefore the decision tree classifier will not provide any substantial increase in efficiency for the classification of LANDSAT data. Also the design of the tree structure for classification of LANDSAT data can be carried out manually with the aid of ellipse plot. An ellipse plot is a graph in which loci of points of constant density of the two dimensional normal density function for all classes are plotted. Therefore the decision tree classifier does not bring any additional information which may help classification. Rather it looks into the feature space with more detail and uses feature subsets in a sequential manner.

You and Fu (88) designed a linear binary tree classifier to take advantage of the accuracy of a decision tree classifier and used linear discriminant functions at decision stages to reduce classification time. A tree classifier is said to be binary if and only if each nonterminal node has only two subtrees. It was found that the linear binary tree classifier with one or two features at each stage is much better in performance and is about twelve times faster in classification speed than the conventional single-stage maximum likelihood classifier with the same number of features, although it takes comparatively more computer time to design the tree.

Another classification procedure called "sample classification" was proposed by Kettig (8). A distinctive characteristic of the spatial dependence in MSS data is "redundance," i.e., the probability of transition from class i to class j is much greater if j=i than if j≠i, because the sampling interval is generally smaller than the size of an object. This suggests the use of an "image partitioning" transformation to delineate the arrays of statistically similar pixels before classifying them. Since each homogeneous array represents a statistical "sample" (a set of observations from a common population), a "sample classifier" could then be used to classify the objects. In this way, the classification of each pixel in the sample is a result of the spectral properties of its neighbors as well as its own. Thus its "context" in the scene is used to provide better classification. ECHO (extraction and classification of homogeneous objects) designates this approach.

Kettig and Landgrebe (63) adopted an image partitioning algorithm due to Rodd (69) to delineate all boundaries in the image. A typical scene is assumed to consist primarily of objects whose boundaries form a partition of the scene. Each object in the partition belongs to one of K classes. If $\underline{X} = (\underline{x}_1, \ldots, \underline{x}_n)$ represents a set of pixels in some object, then this set constitutes a "sample" from a population characterized by one of the class-conditional pdf's. A sample classifier is simply a strategy for deciding which one, based on the n observations. One popular approach is the ML strategy which assigns X to class W_i if

$$\text{Ln } P(\underline{X}/W_i) = \max_j \text{ LnP}(\underline{X}/W_j) \tag{1.2}$$

The advantage of ECHO is the increase in speed and accuracy. Best performance can be achieved by applying this classifier to aircraft data collected at lower altitudes. The aircraft data are characterized by large homogeneous objects in the scene. For the classification of LANDSAT data, however, the ECHO classifier will not be suitable, because there are too many obscure boundaries in LANDSAT data to partition the scene effectively. Also the correlations used are attributable to the effect of dependent states within the same class. Other sources of correlation are ignored in this approach.

The recent literature contains numerous references to image partitioning algorithms. Robertson et.al.

(67) divide an image into two main categories. "Boundary seeking" algorithms characteristically attempt to exploit object contrast. Two of these have been implemented with MSS data (69), but they are incompatible with sample classifiers due mainly to their failure to produce boundaries that always close on themselves. The other category can be called "object seeking" algorithms, which characteristically exploit the internal regularity (homogeneity) of the objects. As the name implies, an object seeking algorithm always produces well-defined samples (and thus closed boundaries as well). There are two opposite approaches to object seeking, which we shall call conjunctive and disjunctive. A conjunctive or merge algorithm begins with a very fine partition and simplifies it by progressively merging adjacent elements together that are found to be similar according to certain statistical criteria (69),(70). A disjunctive or split algorithm begins with a very simple partition and subdivides it until each element satisfies a criterion of homogeneity. For example, Robertson' algorithm (67),(71) is based on the premise that if a region contains a boundary, splitting the region arbitrarily will usually produce two subregions with significantly different statistical characteristics.

Kettig combined Rodd's (69) conjunctive partitioning algorithm with a minimum distance sample classifier and observed an improvement in classification, but processing time was increased (72). Gupta and Wintz (73) added a test of second order statistics to Rodd's first order test, but obtained essentially the same results as the first order test at greater cost in processing time. Robertson (67,71) implemented a disjunctive partitioning algorithm with the same minimum distance classifier. He obtained about the same classification accuracy as conventional no-memory classification with an order of magnitude increase in processing time.

Welch and Salter (1) laid the basic background for contextual pattern classification. They used compound decision theory to introduce contextual information into the decision scheme. However, their basic assumption (made in an effort for a practical solution) was that contextual relationships between nonadjacent cells are negligible. Also, they did not give any theoretical reason as to why compound decision performs better than a no context rule. We will take a detailed look into compound statistical decision theory in Chapter 2.

As we mentioned earlier, few attempts have been made to use contextual information in the classification of imagery data or other kinds of multidimensional data. It can be seen from the above survey that relevant work in the area of using contextual information is at its infancy. Thus, it is our intention to investigate and develop useful contextual classifiers for remote sensing.

1.3 Chapter Outline

In Chapter 2, the compound decision theory and its closely related sequential compound decision, empirical Bayes approach are all discussed. Theoretical justification and intuitive insight are given as to why compound decision rules perform better. Experimental results using derived compound bayes decision rule are also presented in this chapter.

In Chapter 3, the proposed spatial correlation contextual information is exploited by constructing a Markov random field (mesh) on the plane lattice. The way in which this spatial stochastic model (SSM) is fitted on the plane and the extension of this model to the multivariate site variable case are also discussed.

In Chapter 4, a spatial stochastic classification model (SSCM) is introduced. To have a much more efficient contextual classification performance, a recursive SSCM (RSSCM) is developed. Experimental results on LANDSAT data are presented.

Finally, Chapter 5 concludes the whole study. Some detailed derivations and flow charts are placed in Appendices.

Since the application of the contextual classifier to multispectral remote sensing data is emphasized, the assumption of multivariate normal data distributions (which is often a reasonable assumption for remote sensing data) will be constantly used in later derivations.

CHAPTER 2

COMPOUND DECISION APPROACH TO CONTEXTUAL CLASSIFICATION

The first type of context that we would like to employ in making decisions about the unknown patterns is the class transition probability context. For example, there is a nonzero probability that water is next to a boat. Since we are dealing with two dimensional image data, the class transition probabilities are directional. As we will see in this chapter, compound decision theory or sequential compound decision theory is a powerful approach for handling dependent classes. Abend (3) and Raviv (2) point out that a sequential procedure can be implemented fairly efficiently when the classes form a low-order Markov chain. However the prospect is considerably less attractive when they form a Markov mesh (Markov field), which is a more suitable model for two-dimensional scenes. We shall discuss this problem in Chapter 3. Compound decision theory treats all observations (patterns) simultaneously. It does not matter whether the data come from a one-dimensional string or a two-dimensional plane. Therefore, we shall see how we can use all available patterns at hand to make decisions about each one of them.

2.1 Statistical Decision Problems

In the present section we establish the notation to be used throughout. Both simple decision and compound decision problems are discussed. We shall avoid the discussion of measurability questions, and will assume that all formulas given are well defined.

2.1.1 Classical Decision Problem

Consider the following finite statistical problem. Let $\underline{X}$ be a random scalar or vector-valued variable known to have one of the possible distributions P_θ, where (the class) θ belongs to the parameter space Ω, and is unknown. After $\underline{X} = \underline{x}$ has been observed, an action a, belonging to the action space A, must be chosen. The action $a \in A$ is, of course, based on the observed value of $\underline{x}$ and corresponds to the decision made in the pattern recognition problem. If a is chosen, and the true parameter value (class category) is θ, a loss $L(a,\theta)$ is incurred. If the loss function $L(a,\theta)$ is bounded from below, one may without loss of generality assume $L(a,\theta) \geq 0$.

A (statistical) decision rule is a rule which for each observed value $x \varepsilon X$ decides the probability of taking each action of A. Since this action may be chosen from among all possible actions by a random device, the most general description of a decision rule ϕ is ϕ: $X \rightarrow$ {all probability distributions over (A,σ_A)} (σ_A denotes some defined σ - field of subsets of A). Such a ϕ is often called a randomized decision function. The risk function is defined to be the expected loss incurred by ϕ when the true parameter value is θ, i.e.

$$R(\phi,\theta) = \int_{\underline{X}} \int_A L(a,\theta)\,\phi(\underline{x},da)\,dP_\theta(\underline{x}) \tag{2.1}$$

If for each $\underline{x} \varepsilon X$ one action $a_{\underline{x}} \varepsilon A$ is chosen with probability one, ϕ is called a nonrandomized rule. This decision problem is denoted by (Ω, A, L).

The Bayesian approach to such a decision problem assumes that the parameter θ is itself a realization of a random variable with prior distribution G over (Ω,σ_Ω). From the Bayesian point of view, the Bayes risk with respect to G is the overall expected loss. It is given by

$$r(\phi,G) = \int_{\Omega} R(\phi,\theta)dG(\theta). \tag{2.2}$$

A Bayes rule ϕ_G with respect to G is a decision rule which minimizes the Bayes risk (2.2). The minimal value of (2.2) is denoted by $\rho(G)$, i.e.,

$$\rho(G) = \min_{\phi} r(\phi,G) = r(\phi_G,G). \tag{2.3}$$

$\rho(G)$ is called Bayes envelope functional, and is defined over the set of all prior distributions. As long as G is known, a person taking the Bayesian viewpoint has an optimal rule at his disposal, and is not faced with the problem of what rule to select. For the special case where $A = \Omega$ and zero-one loss function, the Bayes risk is equal to the probability of error and the Bayes rule is obviously the best decision rule.

2.1.2 Compound Decision Problem

Suppose the statistical decision problem (Ω,A,L) is repeated n times, where no relationship is assumed among the n repetitions. We can consider the n decision problems having identical generic structure simultaneously. This constitutes a compound decision problem. We are then faced with an unknown vector $\underline{\theta}_n = (\theta_1,\ldots,\theta_n)\in\Omega^n$ (Ω^n denotes the Cartesian products of n sets Ω), and a corresponding vector $\underline{X}_n = (\underline{x}_1,\ldots,\underline{x}_n)$ of random variables (vectors). We observe each random pattern vector $\underline{x}_i$ with distribution P_{θ_i} for each θ_i, and we assume that the $\underline{x}_i$'s are independent. For each θ_i an action a_i must be chosen, the loss incurred being $L(a_i,\theta_i)$.

The remote sensing problem (where we repeatedly classify every image point) can be viewed conveniently in the framework of the theory of compound statistical decision problems. The spectral classifiers which classify each single pixel based solely on data associated with that point form the component problem of the compound problems.

The parameter space of the compound decision problem is Ω^n and the new action space is A^n. The loss function of a compound decision problem is defined as the average of the loss functions of the component problems. That is, for every $\underline{a}_n = (a_1,\ldots,a_n)$ the loss function for the compound decision problem is

$$L_n(\underline{a}_n,\underline{\theta}_n) = \frac{1}{n}\sum_{i=1}^{n} L(a_i,\theta_i). \tag{2.4}$$

Thus the compound problem is characterized by (Ω^n,A^n,L_n).

The problem of developing useful rules for the compound decision problem was neglected until 1951, when Robbins (89) showed that in spite of there being no relation between the n actual problems (i.e. no relation between the unknown parameters $(\theta_1,\ldots,\theta_n)$), one can by means of a compound approach achieve results which, as shown by Samuel (90), can not be achieved by the standard approach. To emphasize that no relation is assumed to hold among the unknown parameters, $\underline{x}_1$ could be an observation on a butterfly in Ecuador, $\underline{x}_2$ on an oyster in Maryland, $\underline{x}_3$ the temperature or a star, and so on, all observations being taken at different times. The standard approach is to decide on each θ by means of $\underline{x}_i$ only, i.e., to treat the component problem (Ω,A,L) n times separately, using ϕ_i for the ith component decision. A rule doing this has been called a simple rule. If the same rule ϕ is used to decide on each of the n components, the corresponding rule is called simple and symmetric.

A compound decision rule is a vector $\underline{\phi}_n = (\phi_1,\ldots,\phi_n)$ where for each given $\underline{X}_n$, ϕ_i specifies the probabilities of taking each of the actions in A for the ith component problem. Therefore ϕ_i is a conditional distribution over A given that the pattern vector $\underline{X}_n$ is observed. We see that the decision functions are now allowed to depend on the data from all n components. The compound risk is taken to be the average of the resulting risks in the component problems. The risk function of $\underline{\phi}_n$, as a function of $\underline{\theta}_n \epsilon \Omega^n$ is according to (2.4) given by

$$R_n(\underline{\phi}_n,\underline{\theta}_n) = \frac{1}{n} \sum_{i=1}^{n} \int_{\underline{X}^n} \int_A L(a,\theta_i)\phi_i(\underline{X}_n,da)\,dP_{\underline{\theta}_n}(\underline{X}_n) \qquad (2.5)$$

where

$$R(\phi_k,\underline{\theta}_n) = \int_{X^n} \int_A L(a,\theta_k)\phi_k(\underline{X}_n,da)\,dP_{\underline{\theta}_n}(\underline{X}_n) \qquad (2.6)$$

is the risk for the kth component problem.

As in the case of the classical statistical decision problem, one can consider the (compound) Bayes risk $r(\underline{\phi}_n,\underline{G}_n)$ with respect to an a priori distribution $G_n(\underline{\theta}_n)$ over Ω^n. The compound Bayes risk $r(\underline{\phi}_n,\underline{G}_n)$ is

$$r(\underline{\phi}_n,\underline{G}_n) = \int_{\underline{\theta}_n \epsilon \Omega^n} R_n(\underline{\phi}_n,\underline{\theta}_n) G_n(\underline{\theta}_n)$$

$$= \frac{1}{n} \int_{\phi_k} \sum_{k=1}^{n} r\,(\phi_k,\underline{G}_n) \qquad (2.7)$$

where

$$r(\phi_k,\underline{G}_n) = \int_{\underline{\theta}_n \epsilon \Omega^n} R(\phi_k,\underline{\theta}_n)\, G_n(\underline{\theta}_n) \qquad (2.8)$$

is the kth component Bayes risk. A procedure is compound Bayes against G_n if it minimizes $r(\underline{\phi}_n,\underline{G}_n)$.

Thus, the compound Bayes procedure $\underline{\phi}_n^G$ is one which for every k, minimizes

$$r(\phi_k,\underline{G}_n) = \int_{X^n}\int_A \sum_{\underline{\theta}_n} L(a,\theta_k)\phi_k(\underline{X}_n,da)G_n(\underline{\theta}_n)dP_{\underline{\theta}_n}(\underline{X}_n) \tag{2.9}$$

Hence ϕ_k should select with probability one that action j which minimizes the quantity

$$\sum_{\underline{\theta}_n} L(j,\theta_k)P(\underline{X}_n,\theta_k), \tag{2.10}$$

where $P(\underline{X}_n,\underline{\theta}_n) = P(\underline{X}_n/\theta_n)G_n(\underline{\theta}_n)$.

If we use a zero-one loss function and action j corresponds to deciding that $\theta = j$, ϕ_k^G chooses the value of θ_k that maximizes $P(\underline{X}_n,\theta_k)$. This is equivalent to maximizing the a posteriori probability

$$P(\theta_k/\underline{X}_n) = \frac{P(\underline{X}_n,\theta_k)}{P(\underline{X}_n)}$$

since the denominator $P(\underline{X}_n)$ is independent of θ_k.

One is usually interested in the problem of finding a criterion by which to judge compound decision rules. This is solved as follows. Consider a hypothetical situation where one is told the actual values of the components θ_i of θ_n, but not their order, before the decision must be made. This is equivalent to saying that one knows the empirical distribution $G_n = G_n(\underline{\theta}_n)$ of the θ-values. Obviously there are at most n! different θ_n vectors whose empirical distribution is G_n. Let $S(G_n)$ denote the set of all such $\underline{\theta}_n$. Robbins (91, Theorem 2) has showed that if one is restricted to simple rules, the simple symmetric rule with components ϕ_{G_n} defined by equation (2.3) is optimal, in the sense that

$$\inf_{(\phi_1,\ldots,\phi_n)} \sup_{\underline{\theta}_n \in S(G_n)} \sum_{i=1}^{n} R(\phi_i,\theta_i)/n = \sum_{i=1}^{n} R(\phi_{G_n},\theta_i)/n.$$

$$= \rho(G_n) \qquad (2.11)$$

Where the infinum is taken over all simple rules, and the equalities hold for every $\underline{\theta}_n \varepsilon S(G_n)$. This gives a new interpretation of the Bayes envelop functional ρ. $\rho(G_n)$ is the minimal attainable expected loss for the compound problem, by a simple symmetric rule, if the empirical distribution of the parameters G_n is known in advance. However it should be noted that, by the definition of a decision rule, no rule $\underline{\phi}_n$ can assume any advanced knowledge about θ_n. Therefore it has become customary to compare the value of the compound risk function (2.5) with the corresponding value $\rho(G_n(\underline{\theta}_n))$, in the hopes that for some rule $\underline{\phi}_n$ the difference between the two will be small for all $\underline{\theta}_n \varepsilon \Omega^n$, at least when n is not too small. The reason for using compound rule instead of some simple rule is that one hopes for large n to be able to extract some information about the prior distribution from the values $\underline{x}_1,\ldots,\underline{x}_n$ which have been observed, hopefully in such a way that the compound rule will be close to the optimal but unknown simple Bayes rule which one would use throughout if one knew the prior distribution G. The validity of such an argument for using compound rules has been shown by Samuel (90). Samuel showed that no simple rule has property (A) as described below:

From the motivation given in (2.11) the word "good" has been given an exact asymptotic meaning, by the following definition.

Definition 2.1.

Let $\underline{\theta} = (\theta_1, \theta_2, \ldots)$ with $\theta_i \varepsilon \Omega$. The sequence $\underline{\phi}_n$ of compound decision rules is said to have property (A), if for every $\varepsilon > 0$ and every $\underline{\theta}$ there exists an integer $N(\underline{\theta}, \varepsilon)$, such that for all $n \geq N(\underline{\theta}, \varepsilon)$,

$$R_n(\underline{\phi}_n, \underline{\theta}_n) - \rho(G_n(\underline{\theta}_n)) < \varepsilon \qquad \text{(A)}$$

Where $\underline{\theta}_n$ is the initial n-vector of $\underline{\theta}$, and $G_n(\underline{\theta}_n)$ is its empirical distribution. $\underline{\phi}_n$ is said to have uniform property (A) if $N(\underline{\theta}, \varepsilon)$ can be chosen to be independent of $\underline{\theta}$.

Let $v(\theta) = \inf_{\phi} R(\phi, \theta)$. If for every θ, $L(a, \theta) = 0$ for some a, then $v(\theta) = 0$. This is true in most cases. The main reason for considering nonsimple compound decision rules is the following theorem, which is given by Samuel (90).

Theorem 2.1.

If for some finite subset M of Ω

$$\inf_{\phi} \sup_{\theta \varepsilon M} (R(\phi, \theta) - v(\theta)) = \alpha > 0 \qquad (2.12)$$

then no sequence of simple rules has property (A).

Proof:

Let (2.12) hold for $M = (\theta_{(1)}, \ldots, \theta_{(k)})$, suppose that the assertion is false. Let $\underline{\theta}^{(j)} = (\theta_{(j)}, \theta_{(j)}, \ldots)$ $j = 1, \ldots, k$, and let $N = \max_{j=1,\ldots,k} N(\underline{\theta}^{(j)}, \varepsilon)$ where

$\varepsilon<\alpha$ and $N(\underline{\theta}^{(j)},\varepsilon)$ is defined by property (A). Let $\underline{\phi}_N = (\phi_1,\ldots,\phi_N)$ be the N-th element of the sequence of simple rules with property (A). Since $\rho(G_N(\underline{\theta}_n^{(j)})) = v(\theta_{(j)})$ we have by property (A)

$$R_N(\underline{\phi}_N,\underline{\theta}_N^{(j)}) - v(\theta_{(j)}) < \varepsilon \text{ for } j=1,\ldots,k \qquad (2.13)$$

Since ϕ_N is a simple rule $R_N(\underline{\phi}_N,\underline{\theta}_N^{(j)}) = \frac{1}{N}\sum_{i=1}^{N} R(\phi_i,\theta_{(j)})$. Thus from (2.13) we have

$$\sup_{\theta\varepsilon M} \sum_{i=1}^{N} R(\phi_i,\theta)/N - v(\theta) < \varepsilon < \alpha$$

which contradicts (2.12).

Therefore for any sequence of simple rules the compound risk will not converge to the minimal attainable risk. Clearly only in trivial cases (2.12) does not hold, and thus we can not be content with considering only simple rules. However this does not give enough evidence that we should use compound rule (use irrelevant data to make its decisions), because we have not shown whether any compound rule has property (A). Fortunately this conjecture is solved by Robbins (92). To understand why compound decision rule is superior than simple rule, we must introduce sequential compound decision rule and empirical Bayes approach.

2.1.3 The Sequential Compound Decision Problem

The compound decision problem (Ω^n,A^n,L_n) is characterized by the fact that all n observations have been obtained before any decision has to be made. This is a practical situation in the remote sensing problem since all image data are at hand before we do any classification. In many other cases, however, the ith decision must be made directly after the ith observation $\underline{x}_i$ has been obtained. Some practical situations fitting the above framework are:

a) Patients arriving sequentially for a medical test, must be classified as sick or healthy
b) In an elaborate manufacturing process, items must be classified into one of several categories, or their quality must be estimated, one by one, after they have been produced.

Decision problem of this type have been termed sequential compound decision problems (93). In the sequential case we can use the observations $\underline{X}_i = (\underline{x}_1,\ldots,\underline{x}_i)$ to decide on the i-th action, $i=1,\ldots,n$. We define a sequential compound decision rule by $\underline{\phi}_n = (\phi_1,\ldots,\phi_n)$ where, for each given $\underline{X}_i$, ϕ_i specifies the probabilities by which the ith action is chosen, $i=1,\ldots,n$. We do not need to introduce any new notation, since every sequential compound decision rule is, by definition, a particular case of a nonsequential compound rule. The risk function (2.5) can be rewritten as

$$R_n(\underline{\phi}_n,\underline{\theta}_n) = \frac{1}{n}\sum_{i=1}^{n} \int_{\underline{X}^n}\int_{A} L(a,\theta_i)\phi_i(\underline{X}_i,da)\,dP_{\theta_i}(\underline{X}_i) \tag{2.14}$$

Again we are interested in the problem of finding rules with property (A). Some hope for the possibility of finding such rules is given by the following simple lemma, a proof of which can be found in Samuel (94, p. 881).

<u>Lemma 2.1.</u>

Let $\underline{\theta}_n$ be any fixed vector, $\underline{\theta}_i$ its initial i-vector, and let $G_i=G_i(\underline{\theta}_i)$. Let ϕ_{G_i} denote any version of a Bayes rule of the component problem, with respect to G_i. Then

$$\frac{1}{n}\sum_{i=1}^{n} R(\phi_{G_i},\theta_i) \le \rho(G_n) .$$

In the search of decision rules with property (A), use of Lemma 2.1 can be made in two directions:

(I) As mentioned before, we like to gain some information about the priori distribution G_i from the values $\underline{x}_1,\ldots,\underline{x}_i$ which have been observed. Thus one may try to estimate G_i by some estimate $\hat{G}_i$ based on $\underline{X}_i$, and then use a simple Bayes rule with respect to $\hat{G}_i$, $\phi_{\hat{G}_i}$, with the ith observation $\underline{x}_i$ to choose the ith action. If the sequence $\hat{G}_i$ is, in some sense, a consistent sequence of estimates of G_i, there may be reason to hope that the risk incurred on the ith step by this compound procedure will be close to $R(\phi_{G_i},\theta_i)$ (the true expected loss, if we knew G_i), for sufficiently large i. If this is indeed the case, then Lemma 2.1 yields property (A).

Samuel (93,95,96) and Van Ryzin (97) have used this method to find sequential compound rules with property (A). We state a well-known definition.

<u>Definition 2.2</u>

Let G be a distribution of (Ω,σ_Ω). The distribution P_G over $(\underline{x},\sigma_x)$ defined by

$$P_G(B) = \int_\Omega P_\theta(B)dG(\theta) \text{ for every } B\varepsilon\sigma_x$$

is called a mixture of the family $P_\theta{:}\theta\varepsilon\Omega$, and G is called a mixing distribution. The mixtures are said to be identifiable if $G\neq G^*$ implies $P_G\neq P_{G^*}$.

In any situation where Ω is finite and the mixtures of the distributions $P_\theta{:}\theta\varepsilon\Omega$ are identifiable, it is easy to obtain estimates $\hat{G}_i$ of G_i such that the maximum difference between the actual and the estimated frequencies of the θ-values tends to zero with probability one. Such estimates have been proposed by Robbins (92, Section 7).

Since the problem of estimating G_i is even more complicated then that of estimating a mixing distribution, method (I) has not been used for problems more general than the finite decision problem.

(II) In some instances the Bayes rule ϕ_{G_i} may be directly estimable by some estimate ϕ_i, based on $\underline{X}_i$, without estimating the prior distribution G_i. In these instances it is more natural to estimate ϕ_{G_i} directly.

The latter approach has been used by Samuel (94), where the component problem considered is an estimation problem, and certain parametric families of distributions are considered. It has also been used in a paper by Johns (98), which considers component problems involving two actions, for certain parametric and nonparametric situations.

In spite of the fact that it is conceptually easy to believe that the above described methods lead to sequential compound rules with property (A), the rigorous proof of this assertion for each of the specific problems considered, have usually been quite long and cumbersome. In order to further gain more evidence why compound (or sequential compound) rules will give us more information about G_i and thus benefits us in making decisions, we will examine the relations between compound rules and empirical Bayes approach. Appendix A gives a derived sequential Bayes decision rule by Raviv (2) and Abend (3) for character recognition problems. Interested readers may refer to Appendix A.

2.1.4 Relation to the Empirical Bayes Approach

The methods of dealing with the compound decision

problem (except the compound Bayes rule) are clearly non-Bayesian. Property (A) is a natural goal from a non-Bayesian point of view.

The above described method is closely related to the methods used in the empirical Bayes approach (100,92). As its name indicates, the empirical Bayes approach is a Bayesian approach, and as such it assumes the existence of a prior distribution G over the parameter space. The parameter is a realization of a random experiment governed by G. The difference between the "classical" Bayesian and the "empirical" Bayesian is that the latter does not assume any knowledge about the prior distribution G. The connection between the compound decision problem and the empirical Bayes approach is explained in a paper by Neyman (99).

An empirical Bayes approach is possible only if the same Bayesian problem is repeated several times, so that there may be some hope to learn something about the prior distribution, G, from the observations. Therefore, the empirical Bayes approach deals with the compound decision problem from a very special Bayesian point of view; the parameter values $\theta_1, \theta_2, \ldots$ are considered to be realizations of independent, identically distributed random variables $\Lambda_1, \Lambda_2, \ldots,$ with the same, but unknown, distribution G. With this approach the conditional distribution of $\underline{x}_i$, given $\Lambda_i = \theta_i$ is $P(\underline{x}_i / \theta_i)$. One will then have a sequence of pairs of random variables (Λ_1, x_1), $(\Lambda_2, x_2), \ldots$ of which only the $\underline{x}_i$'s are observable.

Only the sequential compound decision problem has been treated from the empirical Bayes point of view. Example (a) of section 2.1.3 would fit into the empirical Bayes framework if one is to assume that there is a fixed prior probability for every patient arriving for a medical test to be healthy.

It follows from the above that an empirical Bayes rule is just a sequential compound decision rule. If $G_n(\underline{\theta}_n)$ is the n-fold product distribution G^n, where G is any distribution over (Ω, σ_Ω), the simple symmetric rule with components ϕ_G is a Bayes rule for the (Ω^n, A^n, L_n) problem, and the value of $\rho_n(G^n)$ is $\rho(G)$ (i.e. the value of the envelope function for the component problem, at G). This naturally leads us to the definition that an empirical Bayes rule (i.e. a compound decision rule) $\underline{\phi}$ is said to be "asymptotically optimal" if the Bayes risk incurred on the n-th decision tends to $\rho(G)$, for every G. Mathematically, let

$$r(\phi_n, G) = \int_{\Omega^n} \int_{X^n} L(a, \theta_n) \phi_n(\underline{X}_n, da) dP_{\theta_n}(\underline{X}_n) \, dG^n(\underline{\theta}_n) \quad (2.15)$$

Then $\underline{\phi}_n$ is asymptoticaly optimal if and only if

$$\lim_{n \to \infty} r(\phi_n, G) = \rho(G) \quad (2.16)$$

Clearly (2.16) holds if and only if

$$r_n(\underline{\phi}_n, G^n) = \frac{1}{n} \sum_{i=1}^{n} r(\phi_i, G) \to \rho(G) \quad (2.17)$$

which justifies our consideration of the Bayes risk incurred on the n-th component only, as in (2.16).

It is usually much easier to establish that a rule $\underset{\sim}{\phi}$ is asymptotically optimal, than to prove that it has property (A). This is due to the fact that in the empirical Bayes situation all $\underline{x}_i$'s are independent and identically distributed, and that only the Bayes risk is of concern. Since, for given G the simple rule with components $\phi_G(\underline{x}_i)$ satisfies $r(\phi_G, G) = \rho(G)$, it is in most cases sufficient to find some estimate of ϕ_G, and then to use the Lebesgue dominated convergence theorem to establish asymptotic optimality (see, for example, Robbins (100)). Robbins was the first to consider the empirical Bayes approach, as described above. He (100) has found, under suitable conditions, sequence of decision functions which is asymptotic optimal. The objection raised against the compound decision approach - that it uses irrelevent data to make its decisions - now loses its validity. In the empirical Bayes approach all previous observations are relevant for the present decision, since they entail information about the unknown G. Other works in the area of empirical Bayes approach are by Johns (101,102), Miyasawa (103), Samuel (104), and Robbins (92).

One of the reasons for going into such a detailed description of the empirical Bayes approach is to raise some doubts whether the underlying assumptions of such an approach are always justified in practical situations. If not, it seems safer to consider the compound decision approach, and to seek rules with property (A). If it should turn out that one is fortunate enough actually to be in an empirical Bayes situation, he will not have lost anything, because of the following.

Theorem 2.2.

> If $L(a,\theta)$ is bounded, then every $\underline{\phi}$ with property (A) is an asymptotically optimal empirical Bayes rule for the class of all prior distributions.

The proof of this theorem is given in Samuel (105), and actually the theorem is true also in many instances where $L(a,\theta)$ is not bounded.

The empirical Bayes approach may be useful for the compound decision problem in that it is sometimes not so difficult to find empirical Bayes rules which are asymptotically optimal. It remains then to be proven that they actually also have property (A). This constructive method of finding good compound rules has already been suggested by Robbins (89, p. 147).

We have seen by Theorem 2.1 that no sequence of simple rules have property (A). Furthermore, a very thorough argument as to why compound decision rules (or empirical Bayes rules, or a particular case of sequential compound rules) which have property (A) may be found was given. The exact method to find good compound rules with property (A) is problem dependent and is usually very cumbersome. Nevertheless the belief that compound decision rules will perform better than simple rules is greatly strengthened. Experimental investigations may be boldly carried out in situations where we have the compound decision problem at hand. We will describe some experimental results of using compound Bayes rules as a method of employing contextual information in classification of LANDSAT data in section 2.2. However, we will first complete this section by making some further remarks on the compound decision problem.

2.1.5 Further Remarks on the Compound Decision Problem

(i) The description given above assumed that the family of distributions $P_\theta : \theta \varepsilon \Omega$ is known, and hence the approach was parametric. But by the same token a nonparametric approach is possible. In the nonparametric instances the loss function usually depends upon some nonparametric characteristic of the distribution, such as its expectation, median, variance, etc., and the prior distribution G considered is over the possible values of this nonparametric characteristic. Such a nonparametric approach is considered by Johns (98). It has also been considered for the empirical Bayes situation (101,102).

(ii) It should be emphasized that it is not possible to find rules with property (A) for every conceivable component problem. An example where this is impossible is discussed in Section 5 of Robbins (92), from the empirical Bayesian point of view.

(iii) We have considered only the convergence of the average risk, R_n, in previous discussion. The actual (random) average loss may be of more practical interest. A rule is said to have property (B) a.e. or in probability, if the difference between the average loss and the corresponding value $\rho(G_n(\underline{\theta}_n))$ converges to zero, a.e. or in probability, respectively. Considerations of property (B) can be found in Samuel (93,94,96,105).

(iv) The order of magnitude of the difference $R_n(\underline{\phi}_n, \underline{\theta}_n) - \rho(G_n(\theta_n))$ is a major interest for practical applications. some empirical findings on this problem are given by Samuel (96). For the nonsequential compound problem Hannan and Van Ryzin (106) show that for the component problem of testing a simple hypothesis against a simple alternative, using the rules of Hannan and Robbins (107), the order is always $O(n^{-1/2})$, and sometimes $o(n^{-1})$. Van Ryzin (108) considers the nonsequential compound decision problem for the finite decision problem as component problem. He proposes rules with property (A), and shows that the order of convergence for the rules proposed is $O(n^{-1/2})$ and in some instances $O(n^{-1})$. These results are extended in (97) to the sequential situation. Considerations about the rate of convergence can also be found in (98), for the problem treated there.

(v) The goal we have described for a compound decision rule is to obtain property (A), i.e. we compared the average risk of our rule at $\underline{\theta}_n$ with the value $\rho(G_n(\underline{\theta}_n))$.

The question may be asked whether this goal is optimal in some sense. Swain (109) compares the average risk of some rules for the sequential compound estimation problem with other "objective functions," of which $\rho(G_n(\underline{\theta}_n))$ is a particular case. These objective functions, which we denote $U_k(\underline{\theta}_n)$, $k=1,2,\ldots$, are defined similarly to $\rho(G_n(\underline{\theta}_n))$, except that a k-dimensional empirical distribution of $\underline{\theta}_n$ is considered, rather than the usual one dimensional distribution, and thus the usual envelope function must be replaced by a corresponding envelope function ρ_k. Thus $U_1(\underline{\theta}_n) = \rho(G_n(\underline{\theta}_n))$. Swain shows that for the problem considered

$$U_{k+1}(\underset{\sim}{\theta}_n)<U_k(\underset{\sim}{\theta}_n)+O(n^{-1}),\ k=1,2,\ldots \tag{2.18}$$

and thus the requirement that for all $\underset{\sim}{\theta}$

$$R_n(\underset{\sim}{\phi}_n,\underset{\sim}{\theta}_n)<U_k(\underset{\sim}{\theta}_n)+\varepsilon \quad k>1 \tag{2.19}$$

or sufficiently large n, seems more stringent than property (A). Rules satisfying (2.19) are exhibited in (109). Johns (98) also considers some rules for the two-action problem, which satisfy (2.19), and he establishes that (2.18) holds also for this problem.

In the opinion of Samuel (94) unless there is reason to believe that the actual parameter values $\theta_1,\theta_2,\ldots$ are related in a very special way, one should be content with $U_1(\underset{\sim}{\theta}_n)$, i.e. with property (A). The additional complication of the rules which achieve (2.19) raises doubts about their practicality for applications. If a particular relationship between the parameter values can be assumed, an extended empirical Bayes approach corresponding, e.g., to the generalization of Barndorff-Nielson (110), may be more proper.

After we have given such detailed exposition of compound decision problem, practical applications to the area of remote sensing is of particular interest. This is because of the fact that repeated classification of each pixel in remote sensing fits into the framework of the compound decision problem perfectly. It seems that the performance in classification accuracy using any suitable compound decision rule will be better than that of the classical simple decision rule. In the next section we will describe an experimental investigation on the contextual classification of MSS data using the compound decision approach.

2.2 Compound Decision Approach-Experimental Results

In this section we describe our experimental procedures and results of employing a compound decision algorithm in the classification of remotely sensed data. At the end of this section we will examine the relative advantages and disadvantages of this approach. This will lead us to consider another approach of using contextual information which is discussed in detail in Chapter 3.

2.2.1 Compound Decision Rules

To demonstrate the value of contextual information in classifying LANDSAT data, the compound Bayes decision rule is used as a preliminary approach to introduce context into the classification model. The parameter space and the action space are identical and they represent the list of image truth categories into which we are to classify each cell or pixel of the image. Every image truth category is statistically characterized by a multivariate probability density function. The zero-one loss function is used such that the risk becomes the probability of misclassification.

From the classical statistical decision theory it can be shown that a decision rule which minimizes the simple Bayes risk is to choose the action which minimizes the sum

$$\sum_{\theta} L(\theta, a)P(x/\theta)G(\theta) \tag{2.20}$$

Where $G(\theta)$ is the prior probability of occurrence of class θ. This rule provides the optimum strategy for a classification system not using context (1)

If we have a prior distribution on the parameter space, the decision rule which minimizes the compound Bayes risk chooses the action $j \varepsilon A$, for each individual image cell k, which minimizes the expression (see also equation (2.10))

$$\sum_{\theta_k} L(\theta_k, j)P(\underline{X}_n/\theta_k)G(\theta_k) \tag{2.21}$$

where $\underline{X}_n = (\underline{x}_1, \ldots, \underline{x}_n)$ is an assemble of pattern vectors for all cells in the image frame. Although the problem is defined on the entire frame, the optimum decision rule makes an individual decision on each cell. The compound decision differs from the simple decision only in that it uses all of the pattern vectors from the frame to estimate the class θ_k, while the simple decision uses the vector from the cell k only (1).

Although the minimization of (2.21) provides a general rule for employing context, the direct calculation of the expression is impossible in almost all practical situations. For example, in our simulation, if we limit ourselves to the consideration of one imagery frame at a time, $n=128^2$ and $\underline{X}_n$ is a 128^2-dimensional vector of vectors. The expression $P(\underline{X}_n/\theta_k)$ is a density function defined on this vector space and cannot be evaluated without making some simplifying assumptions. The simplification is also motivated by the following observation. Two pixels in the image are unconditionally correlated, with the degree of correlation decreasing as the distance between them increases. Therefore we would expect that only the neighboring pattern vectors to, say, cell k will entail information about the unknown class θ_k. This suggests us to use some kind of neighborhood definition such that only the pattern vectors which are neighbors of, say, cell k are used in evaluating the joint density function (2.21) to classify cell k. The dimensionality of the joint density function will be immediately reduced considerably. Welch and Salter (1) have arrived at a 4-neighbor implementable context decision rule using this idea. We adopted their assumptions to simplify $P(\underline{X}_n/\theta_k)$ and arrived at an 8-neighbor decision rule. The assumptions used will be discussed in this section. The derivation of the decision rules is given in Appendix B.

	k_2	
k_3	k	k_1
	k_4	

Figure 2.1 Labeling of neighbors of cell k.

Defining adjacent cells as any two cells with a common side, the first assumption is that contextual relations between nonadjacent cells are negligible. We implement this assumption by stating that for any nonadjacent cells, say, b and c,

$$P(\underline{x}_b/\theta_c, \text{ and other } x \text{ or } \theta) = P(\underline{x}_b/\text{and other } x \text{ or } \theta) \quad (2.22)$$

Note that this assumption is not logically sound. According to this assumption, cells k_1 and k_4 have no contextual relation while either one is related with cell k (See Figure 2.1). Nevertheless we shall adopt this assumption since it is mainly intended to enable us to evaluate the joint density function. A much more consistent neighborhood definition will be given in Chapter 3.

The second assumption is that the patter $\underline{x}_k$ of a cell θ_k is a function only of θ_k, the true class of the cell, and if θ_k is known, neither the class nor the pattern of any other cell provides additional information about $\underline{x}_k$. Symbolically, this is

$$P(\underline{x}_k/\theta_k, \text{ any other } \theta \text{ or } \underline{x}) = P(\underline{x}_k/\theta_k) \quad (2.23)$$

This assumption is equivalent to saying that other pattern vectors will not affect the present pattern being classified.

With the above two assumptions, the final 4-neighbor decision rule is to choose action $a \epsilon A$ to minimize

$$\sum_{\theta_k} L(\theta_k, a) P(\underline{x}_k/\theta_k) G(\theta_k) \prod_{i=1}^{4} \sum_{\theta_{k_i}} P(\underline{x}_{k_i}/\theta_{k_i}) P(\theta_{k_i}/\theta_k) \quad (2.24)$$

Notice that the part of the expression outside of

the product sign is identical to (2.20) for the simple decision rule. The product term represents the contextual information for cell k. The four multipliers in the product term represent the contextual contributions from the four adjacent neighbor cells.

If we further assume that the contextual information from the "second order" neighboring cells, namely, k_5, k_6, k_7, k_8 (Figure 2.2) is independent of that from the first order neighbors, we can arrive at an 8-neighbor decision rule. the final form of the 8-neighbor decision rule is to choose $a \varepsilon \underline{A}$ to minimize

$$\sum_{\theta_k} L(\theta_k, a) P(\underline{x}_k/\theta_k) G(\theta_k) \prod_{i=1} \sum_{\theta_{k_i}} P(\underline{x}_{k_i}/\theta_{k_i}) P(\theta_{k_i}/\theta_k) \cdot P($$

$$\cdot \prod_{i=5}^{8} \sum_{\theta_{k_i}} P(\underline{x}_{k_i}/\theta_{k_i}) P(\theta_{k_i}/\theta_k) \qquad (2.25)$$

k_7	k_2	k_6
k_3	k	k_1
k_8	k_4	k_5

Figure 2.2 Second Order Neighborhood of Cell k.

2.2.2 Data Processing

The launch of the first Earth Resources Technology Satellite (ERTS), now called LANDSAT, provided a new perspective for remote sensing research. Experiments were performed on this multispectral scanner data. The Tippecanoe County (Lafayette), Indiana subframe registered under LARS Run No. 72053609 was selected as the study area. Four bands of digitized, multispectral data from the ERTS pass of 30 September 1972 were analyzed by computer. Figure 2.3 shows three different bands of this digitized data. Pattern recognition techniques can be used to process the random data (13), (12), (17). We shall

Band 0.50-0.59 μM

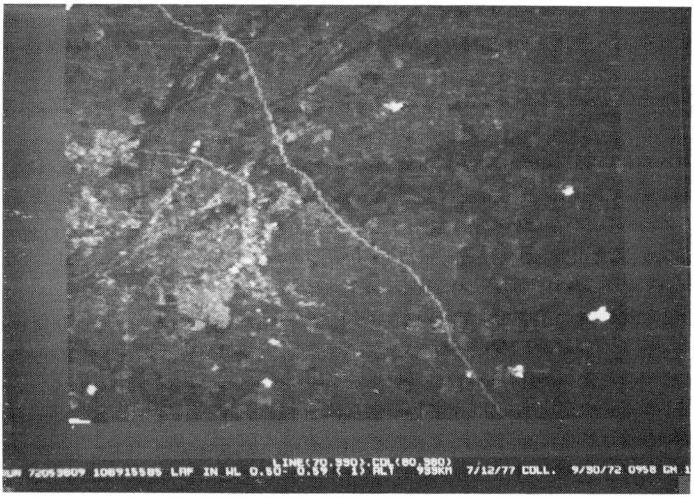

Band 0.60-0.69 μM

Figure 2.3. Three bands of Lafayette, Ind. MSS data.

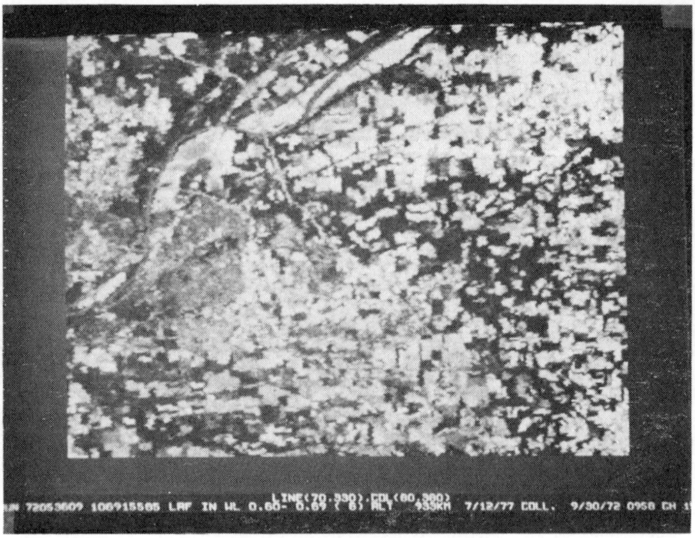

Band 0.70-0.79 μM

Figure 2.3 (continued)

briefly describe the data analysis procedures. For a detailed discussion on LANDSAT data processing, see (17).

The Tippecanoe county data were initially viewed on a digital imaging display for the purpose of orientation. At the same time, several small areas were defined for the clustering algorithm. Selected training samples of all four bands were grouped on the basis of inherent properties within the data. These natural groupings in the data are called cluster classes. There is a tendency for the data points within each cluster class to be distributed in a Gaussian fashion. In many other cases, however, the distribution will exhibit multi-modal (non-Gaussian) phenomenon. The multi-modal density function could be decomposed into Gaussian components by doing clustering again. These components are commonly referred to as subclasses. In our experiment the number of training samples used to train all seventeen cluster classes (including subclasses) is 3372.

Next step in the analysis is to obtain statistical

characteristics of those cluster classes. The purpose of this step is trifold. First it is necessary to have the mean vector and covariance matrix to represent each class. Secondly, the number of clusters available at this point is normally greater than the number of classes needed to train the classifier. We would like to reduce the number of training classes as much as we can, because a smaller number of classes saves computation time and simplifies the interpretation of results. Thirdly, we would like to have some indication of the probability of correct classification in advance of doing any classification. If there appears to be confusion between classes, more clustering can be done on the area already used, or we might choose to go back and select additional candidate training areas in an effort to obtain good distinction between classes.

To determine which cluster classes should be combined and to get an indication of probability of correct classification, a distance measure, called transformed divergence, is used to determine the separability between classes. The reason why transformed divergence was used is that it correlates better with the probability of correct classification (in particular, in a saturating fashion (16)).

Figure 2.4 shows experimental results of plotting probability of correct classification versus transformed divergence (16). This plot can be used as a guide to indicate whether we can combine spectrally similar classes once we select a distance threshold as well as the probability of correct classification. The threshold used in the experiment was 1.0 and the maximum correct classification accuracy that can be obtained was about 84%.

2.2.3 Estimation of Transition Probabilities

One important step in this experiment is the estimation of directional transition probabilities. In order to obtain these probabilities, ground truth information is required to do the estimation. The ground truth data were obtained from a detailed four-band maximum likelihood classification result performed by a LARS professional photo analyst.

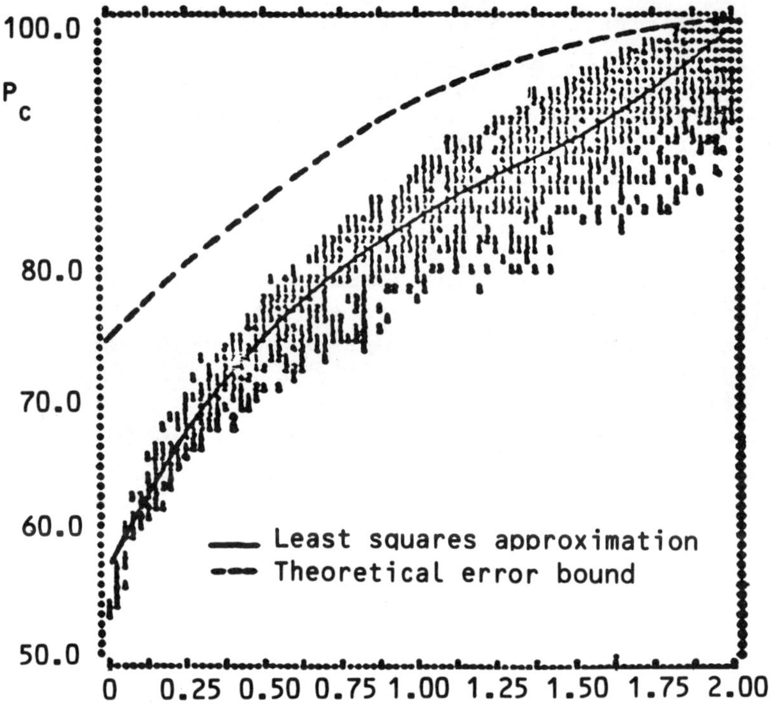

Contextual information is learned from these directional transition probabilities. It was felt that the ground truth obtained by the data analyst was reliable enough to estimate the transition probabilities.

If the classes are denoted by 1,2,..,r, then for a sample size N, the maximum likelihood estimate of p(i), the a priori probability of class i, is n_i/N where n_i is the number of occurrences of class i. The estimate of the transition probability $P(\theta_{k_i} = j/\theta_k = i)$ would be n_{ij}/n_i where n_{ij} is the number of occurrence of θ_{k_i} being j with θ_k being class i. Since the whole data frame is of size 379 lines x 494 columns, we have used a sizable portion (280 lines x 270 columns) of this set in order to obtain "reliable" estimation of transition probabilities.

2.2.4 Experimental Results of Compound Decision Approach

The cospectral plot, as shown in Figure 2.5, was obtained for seventeen classes and for all four bands. It was observed that most separability information was provided by band 0.7-0.8 μM and 0.6-0.7 μM. This fact was later verified by comparing the results obtained by using all four bands to that by using only two bands. These two bands were used to form a bivariate normal density function for each class. The estimated prior probabilities are given in Table 2.1.

The compound Bayes decision rule is employed to classify each pixel into one of a candidate set of classes based on the multispectral data from the pixel itself and the data from either the neighboring four or eight pixels. Figure 2.6 shows the results of applying this approach for classifying a set of LANDSAT MSS data. The block of imagery of size 128 x 128 (lines 200-327, columns 120-247) was classified using the simple decision rule (no context), with the 4-neighbor rule, and with the 8-neighbor rule. Samples of 900 pixels (30 x 30) from this block area were selected exhibiting a range of pointwise accuracies and the corresponding accuracies obtained by using the context-incorporating rules were tabulated together with the corresponding classification time used by IBM 360 computer. The results of the experiment show, as expected, that the classifier using context is consistently better than the classifier which makes each decision based on data from the individual pixels. Furthermore, the "8-neighbor classifier" is consistently better than the "4-neighbor classifier." The final classified images for the simple decision rule, 4-neighbor rule, and 8-neighbor rule and the ground truth information (4-band MLC result by photo analyst at LARS) are shown in Figure 2.7.

The seven information classes used for evaluating the classification accuracies are highway, agricul-

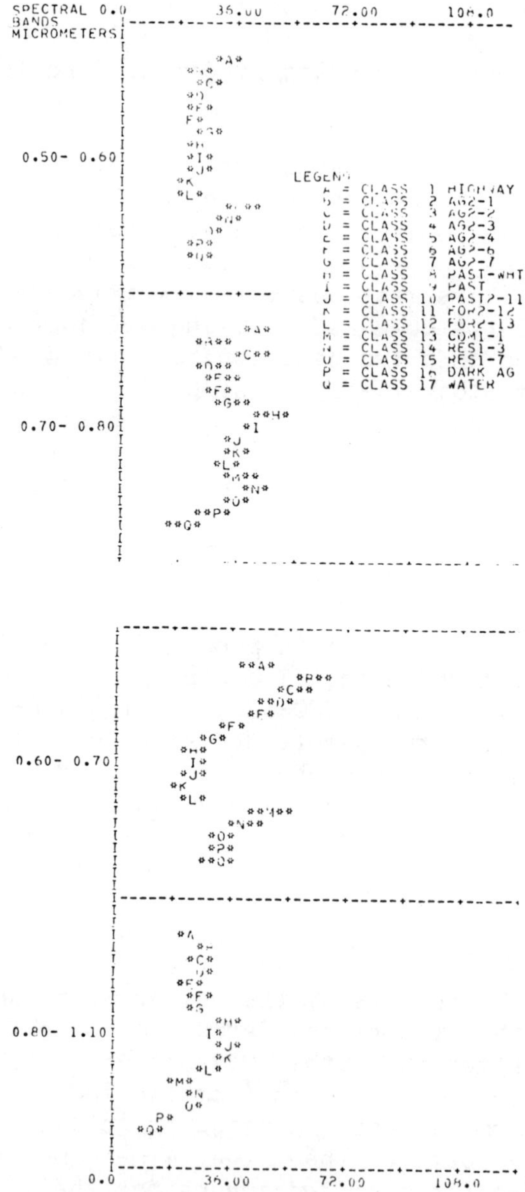

Figure 2.5. 4-band cospectral plot for 17 classes.

Class	Prior Probabilities
1 (Highway)	0.004490
2 (AG2-1)	0.066088
3 (AG2-2)	0.106341
4 (AG2-3)	0.065395
5 (AG2-4)	0.152669
6 (AG2-6)	0.067301
7 (AG2-7)	0.052684
8 (PAST-WHT)	0.032485
9 (PAST)	0.076348
10 (PAST2-11)	0.024450
11 (FOR2-12)	0.072284
12 (FOR 2-13)	0.102384
13 (COM1-1)	0.017148
14 (RES1-3)	0.035336
15 (RES1-7)	0.046342
16 (DARK AG)	0.066688
17 (WATER)	0.011565

Table 2.1 Estimated prior probabilities for17classes.

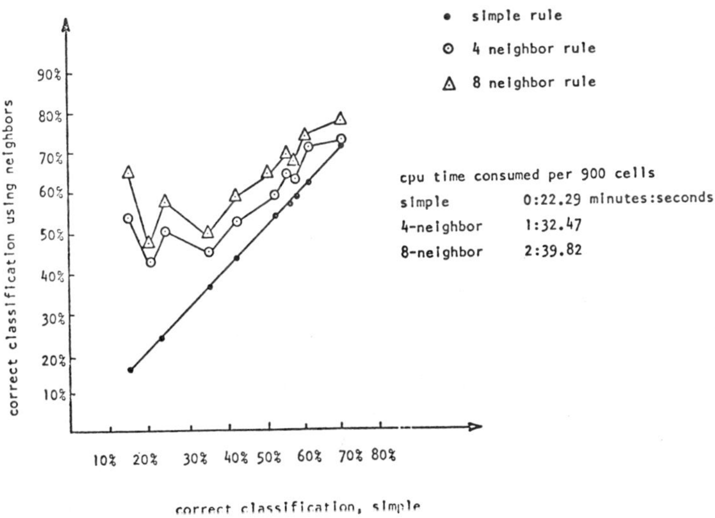

Figure 2.6. Results of an experiment in utilizing compound decision rule.

a) MLC result

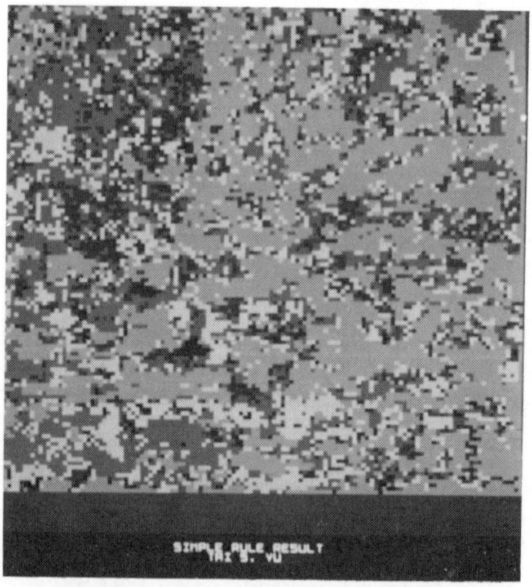

b) Simple Rule result

Figure 2.7. Compound decision rule results of area block (line 200-327, column 120-247).

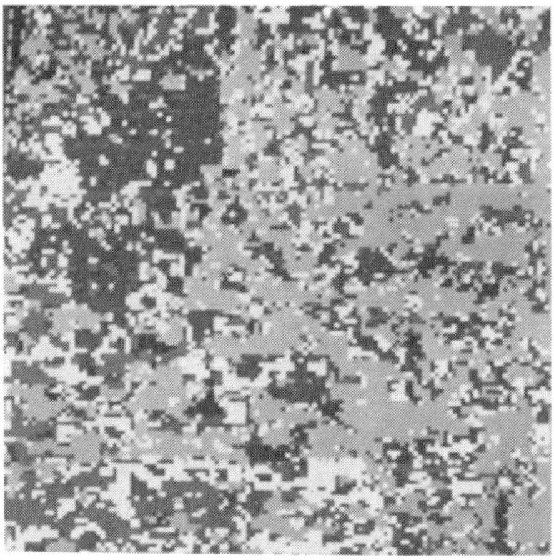

c) 4-neighbor rule result

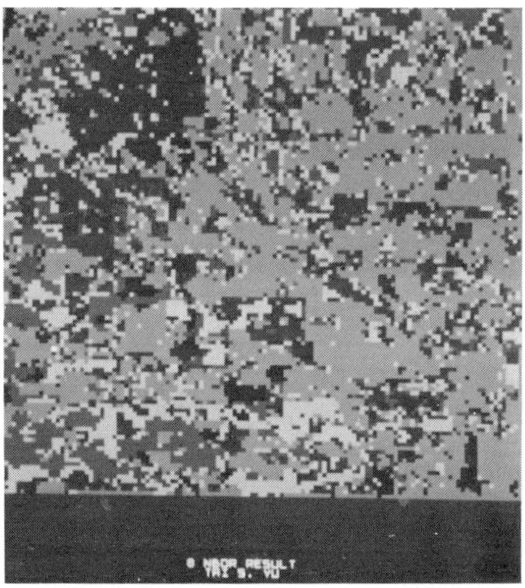

d) 8-neighbor rule result

Figure 2.7 (continued)

ture area, pasture, forest, residential area, commercial area and water. The overall performance of this area block for these seven classes is 56.3% for the simple decision rule, 64.8% for the 4 neighbor rule and 70% for the 8-neighbor rule. It has been found experimentally that for this data set there is a 7% difference in using a 2-band classification compared to that of 4-band. The 14% improvement in classification accuracy provided by the 8-neighbor rule is thus quite substantial.

The price paid for using contextual information is the increased computation time. Different computation times are needed for different decision rules that are derived under different assumptions. The advantage of the assumptions made is that they give explicit decision rules and decision thus make the compound decision procedure feasible in real situations. However justifications of these assumptions has yet to be made. In our case these assumptions considerably limit the kind of context that it is convenient to use. It was assumed that the neighboring patterns themselves do not contribute any information. In other words, "appearance modifier" context (1) is not used here. also the derivations of the compound decision rules will be severely handicapped if we go beyond the eight neighbors. This is because of the fact that in the derivations advantage was taken of the geometric orientation of the four or eight neighbor system. We will summarize these thoughts and make some other remarks before we describe a new approach of using contextual information in the next chapter.

2.3 Concluding Remarks

Even though the compound decision approach provides one way of using contextual information, there are several disadvantages and difficulties when the approach is applied.

(1) The contextual information used is represented in the form of directional transition probabilities. For example, highways occur with large probability next to green vegetation than water. Therefore it is necessary to know the true classes of the neighbors before context

information can be used. However, only the neighboring patterns (and not the true classes) are usually available to us. This means that we must somehow first classify the neighbors before contextual classification of a given pattern can be performed. Thus, how well the context can be used partly depends upon the accuracy with which the preclassified neighbors are classified. The classification process is now clearly going in circles. The reason for this confused compound strategy is that only the presence indicator (1) (transition probability) type of context is used in this approach. A much more realistic type of context is the correlation (5) (appearance modifier (1)) among random variables associated with pixels. We will see in the next chapter how the correlations among pixels can be extracted.

(2) The second problem with compound approach is the reliability of the estimated directional transition probabilities. It is known that an enormous amount of ground truth data must be required in order to have a reliable estimate of such probabilities. However in real situations we seldom have as much ground truth as we would desire to have. The result is that contextual information sometimes is erroneously used. For example, the estimated probability that automobiles are next to highways from limited ground truth data is much lower than that we would have obtained had we have all the ground truth in the whole United States.

(3) The compound decision rule claims that the contextual contributions from the four nearest neighbors are independent, as can be seen from the product term in (2.24). However the occurrence of one class at one pixel will undoubtedly affect the occurrences of its neighboring classes. This fact can be seen from Figure 2.8 where we have pasture present on both sides of a highway. Clearly the occurrence of cell 3 being highway will suggest that cell 5 also is highway and cell 2 being pasture suggests that cell 4 is also pasture.

In other words the contextual contributions from four neighboring cells are hardly independent.

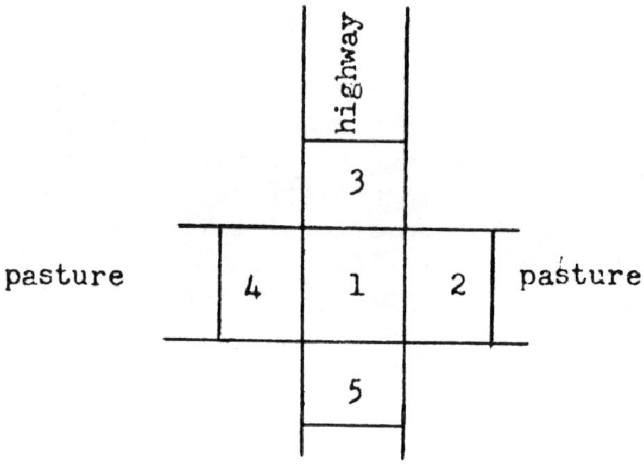

Figure 2.8. A LANDSAT data configuration.

(4) The definition of neighborhood used in the compound decision approach is far from perfect as we mentioned earlier. It would be extremely hard to derive practical compound decision rules using contextual neighborhoods of arbitrary shape. We shall give a much better neighborhood definition in the next chapter. Another result of adopting the neighborhood definition in (1) is the resulting computational complexity of the compound decision rule. If there are r classes and a two-dimensional joint density function is used, an additional {4[(26+1) r]+5] r elementary operations are required to evaluate the decision rule. Reducing computation time thus becomes another goal of our proposed approach.

CHAPTER 3

SPATIAL STOCHASTIC MODEL FOR CONTEXTUAL CLASSIFICATION

In this chapter we will discuss a stochastic model for spatially oriented points (sites) in which correlations exist between any pair of sites. In particular, emphasis will be placed upon two-dimensional lattice points. To see why such a model is needed, we will first give some motivation for the proposed approach.

3.1 Motivation and Proposed Approach

The kind of context we would like to exploit is the correlations existing between the "site-variables" on the two-dimensional lattice. Every pixel in the LANDSAT MSS data can be considered as having one random variable associated with it. Two pixels in spatial proximity to one another are unconditionally correlated, with the degree of correlation decreasing as the distance between them increases. The best way to incorporate these correlations statistically is to consider the joint probability density function of all the site-variables involved. For example, a five dimensional joint probability density function will account for all correlations between any pair of sites in the 4-neighbor system (Figure 2.1). Similarly a nine dimensional joint density function is used for the eight-neighbor system (Figure 2.2). Clearly it is desirable to be able to write down the joint density function of any subset of all site-variables on the lattice. When the joint density functions of all subsets of random variables are specified consistently, we say that a stochastic process has been defined on those random variables. If the random variables are located in two or more dimensional space, the consistently defined process will be called a spatial stochastic process, or a random field.

Since all the spatial correlations among site-variables on a lattice can be extracted by the specified spatial process, we would like to see how they may be useful in the classification process. The most important quantity in evaluating the discriminant function is the (joint) probability density function. Clearly only the joint density function of all the site-variables within the specified contextual neighborhood (window) needs to be considered. For example, in the 4-neighbor system, we would consider the joint probability function $P(x_k, x_{k_1}, x_{k_2}, x_{k_3}, x_{k_4})$ in evaluating the discriminant function. The joint density function $P(x_k, x_{k_1}, x_{k_2}, x_{k_3}, x_{k_4})$ can be decomposed as the product of $P(x_k/x_{k_1}, x_{k_2}, x_{k_3}, x_{k_4})$ and $P(x_{k_1}, x_{k_2}, x_{k_3}, x_{k_4})$. If the site variable x_k is indeed dependent upon its four neighbors, the conditional distribution $P(x_k/x_{k_1}, x_{k_2}, x_{k_3}, x_{k_4})$ ought to be different from the unconditional distribution $P(x_k)$. Therefore an immediate task in our investigation is to find out the functional form of the conditional probability function $P(x_k/x_{k_1}, x_{k_2}, x_{k_3}, x_{k_4})$ such that the consideration of the joint density function $P(x_k, x_{k_1}, x_{k_2}, x_{k_3}, x_{k_4})$ in the classification stage may not be meaningless.

In the following section the conditional probability function is specified in defining a spatial stochastic process. The explicit functional form is also known for the Gaussian case. We will summarize several advantages of the proposed approach below before going into next section.

(1) Unlike the compound decision approach, the proposed approach extracts the real contextual information from the neighboring patterns, rather than from ground truth. Our goal is to find the correlations between sites purely from the data themselves without relying heavily on ground truth data.

(2) Because of the processing of extra information, the computation time for any contextual algorithm will inevitably be increased. It is usually a practical goal to reduce the processing time as much as possible. The compound decision rule (2.24) undoubtly is not the most computationally efficient context algorithm, because many extraneous computations have to be performed to approximate the joint probability function. In the proposed context algorithm, however, the joint probability function can be evaluated directly, resulting in a potential increase in speed (decrease in cost).

(3) The proposed approach is general. If we desire to change the contextual neighborhood for the classification process, the most apparent change in the decision rule occurs in the joint density function. Since a stochastic process has been defined on the data points, we can easily evaluate the joint probability function of any contextual neighborhood points. This means that the decision rule for the new contextual neighborhood will not need to be rederived, an advantage which the compound decision approach failed to accomplish.

These three advantages of the proposed approach are also the goals which any contextual classification algorithm would like to accomplish. Realization of these advantages will be demonstrated through the experimental results given in Chapter 4. In the following sections we will discuss the construction of stochastic processes on the plane lattice points.

3.2 Specification of Spatial Stochastic Processes

In this section we will describe one method of constructing a spatial process for points on a two dimensional lattice. Later in the discussion the method is generalized to include points in multidimensional space. Since LANDSAT data are displayed on finite sized rectangular lattices, the rectangular lattice process will be developed in this section. However, we will find out later that the spatial correlations between pairs of points are not position-independent (stationary). This complicates the classification process considerably, because the correlations will have to be repeatedly calculated as we classify different points in the lattice. Nevertheless, with some approximations, it is possible to implement this rectangular lattice model to perform contextual classifications. Our eventual goal, however, is to develop a stationary lattice process. The stationary process can be constructed after we have constructed a rectangular process. This procedure is discussed in Section 3.3. In the next few sections we will concentrate our attention on the rectangular lattice.

3.2.1 Conditional Probability Approach to Spatial Processes

Whittle (30) and Bartlett (31) gave two non-equivalent definitions of nearest-neighbor systems. If one has a rectangular lattice with sites labeled by integer pairs (i,j) and with an associated set of random variables (X_{ij}), then Whittle's definition requires that the joint probability distribution of the variates should be of the product form

$$\pi_{ij}\ Q_{ij}(x_{i,j},x_{i-1,j},x_{i+1,j},x_{i,j-1},x_{i,j+1}) \qquad (3.1)$$

Where x_{ij} is a value of the random variable $X_{i,j}$. On the other hand, Bartlett's definition requires that the conditional probability distribution of $x_{i,j}$, giving all other site values, should depend only upon the values at the four nearest sites to (i,j). Mathematically, it is

$$P(x_{i,j}/\text{all other site values}) =$$

$$P(x_{i,j}/x_{i-1,j}, x_{i+1,j}, x_{i,j-1}, x_{i,j+1}) \qquad (3.2)$$

The conditional probability formulation apparently has more intuitive appeal. However, it is marred by a number of disadvantages.

First, there is no obvious method of deducing the joint probability structure associated with a given conditional probability model. Second, the conditional probability structure itself is subject to some awkward and highly restrictive consistency conditions. Third, it has been remarked by Whittle (42) that the natural specification of an equilibrium process in statistical mechanics is in terms of the joint distribution rather than the conditional distribution of the variables.

Had these conditional probability models failed to retain any practical appeal, there would have been little reason to pursue them further. However, this is not the case. Suppose one considers wider classes of conditional probability models in which the conditional distribution of $x_{i,j}$ is allowed to depend upon the values at more remote sites. One can build up a hierarchy of more and more general models. That is, one extends the concept of first, second, and higher-order Markov chains in one dimension to the realm of spatial processes. This is also the reason why this proposed approach is general. In the compound decision approach, it is impossible to build up any general higher order models.

The first problem we need to examine is how to derive the joint probability structure associated with the site variables, given their individual conditional distributions. To see why the joint probability structure is required, we will briefly review one-dimensional stochastic processes. It is also interesting to see how a one-dimensional process is constructed since it draws a strong analogy to the spatial (multi-dimensional) process construction.

A stochastic process with a discrete time parameter consists of an infinite sequence of random variables. The process involves a probability measure defined on this infinite-dimensional space. We are particularly interested in probabilities associated with a finite number of the random variables. such probabilities include the cumulative distribution function $\Pr(y_1 \leq b_1, \ldots y_T \leq b_T)$ and probabilities derived from it. Therefore, we want the class of measurable sets to include the sets $(y_1 \leq b_1, \ldots y_T \leq b_T)$; these sets and other sets in the class which are defined in terms of a finite number of coordinates, $y_{t_1}, \ldots y_{t_n}$, are called cylinder sets. Thus the probability measure of the stochastic process implies the probabilities of cylinder sets, including the cumulative distribution functions.

Conversely, if cummulative distribution functions for all finite sets of t's are given and are consistent, then a probability measure can be defined on the infinite-dimensional space so that the probability measure of cylinder sets agrees with the probability as computed from the given cdf's; this extension is unique. That the family of cdf's is consistent means that (i) any marginal (joint) cdf derived from a given cdf corresponds to the given marginal cdf, and that (ii) the cdf's for the arguments written in different orders agree. More precisely, let $F(a_1, \ldots a_n; t_1, \ldots t_n)$ denote a cdf at $a_1, \ldots a_n$; suppose this is given for every real n-tuple $a_1, \ldots a_n$, every $t_1, \ldots t_n$ (from ..., -1, 0, 1, ...), and n=1,2,... . The requirements are

$$(1)\quad F(a_1, \ldots a_m, \infty, \ldots \infty;\ t_1, \ldots t_m,\ t_{m+1}, \ldots, t_n)$$

$$= F(a_1, \ldots a_m;\ t_1, \ldots t_m),\ m = 1, \ldots, n$$

$$(2)\, F(a_1, \ldots a_n;\ t_1, \ldots, t_n) = F(a_{i_1}, \ldots a_{i_n};\ t_{i_1}, \ldots t_{i_n}),$$

where $(i_1, \ldots, i_n)$ is any permutation of $(1, \ldots, n)$.

Then $F(a_1,\ldots,a_n;\ t_1,\ldots t_n)$ may be interpreted as the cdf of $y_{t_1},\ldots,y_{t_n}$. This requirement was given by Kolmogorov (52).

Therefore, in the specification of the spatial process, one must find the joint distributions of all the site variables of the rectangular lattice in order to know all the joint probabilities for any finite set of these site variables. Since a conditional probability function is used for the specification, a link between the joint probability structure and the given conditional density function must be found.

Besag (37) no longer restricts attention to nearest-neighbor (NB) system nor events to lattice schemes in developing the theory. But instead he considers an arbitrary system of multidimensional points (sites) which have spatial interactions existing among them. Suppose that one has a finite set of random variables, $x_1,\ldots,x_n$, which are associated with sites labeled $1,\ldots,n$, respectively. For each site, $P(x_i/x_1,\ldots,x_{i-1},\ x_{i+1},\ldots x_n)$ is specified and one requires the joint probability distribution of the variables. The following assumption is made: if $x_1,\ldots x_n$ can individually occur at the sites $1,\ldots,n$, respectively, then they can occur together. Formally, if $P(x_i)>o$ for each i, then $P(x_1,\ldots x_n)>o$. This is the positivity condition by Hammersley and Clifford (27). Define the sample space $\underline{\Omega}$ to be the set of all possible realizations $\underline{X} = (x_1,\ldots x_n)$ of the system. That is, $\underline{\Omega} = (\underline{X}:P(\underline{X})>0)$. It then follows that for any two given realizations $\underline{X}$ and $\underline{Y}\varepsilon\underline{\Omega}$.

$$\frac{P(\underline{X})}{P(\underline{Y})} = \prod_{i=1}^{n} \frac{P(x_i/x_1,\ldots x_{i-1},y_{i+1},\ldots,y_n)}{P(y_i/x_1,\ldots x_{i-1},y_{i+1},\ldots,Y_n)} \tag{3.3}$$

The proof of this result is given by Besag (23), (37). It is given briefly below.

One may write

$$P(\underline{X}) = P(x_n/x_1,\ldots x_{n-1})P(x_1,\ldots,x_{n-1}) \tag{3.4}$$

However, $P(x_1,\ldots x_{n-1})$ can not be factored in a way similar to (3.4) since $P(x_{n-1}/x_1,\ldots,X_{n-2})$ is not given.

Nevertheless, one can introduce y_n, write

$$P(\underline{X}) = \frac{P(x_n/x_1,\ldots,x_{n-1})}{P(y_n/x_1,\ldots,x_{n-1})} P(x_1,\ldots x_{n-1},y_n)$$

and now operate on x_{n-1} in $P(x_1,\ldots x_{n-1},y_n)$. This yields

$$P(x_1\ldots x_{n-1},y_n) = \frac{P(x_{n-1}/x_1,\ldots x_{n-2},y_n)}{P(y_{n-1}/x_1,\ldots x_{n-2},y_n)}$$

$$P(x_1,\ldots x_{n-2},y_{n-1},y_n)$$

after a similar introduction of y_{n-1}. Continuing this reduction process, one finally arrives at equation (3.3) which clearly determines the joint probability structure of the system in terms of the given conditional probability distributions (37).

Equation (3.3) reveals two fundamental difficulties with the conditional probability approach to specification of the system (37). First, since the labeling of individual sites in the system is arbitrary, this implies that many factorizations of $P(\underline{X})/P(\underline{Y})$ are possible. All of these must be equivalent. This further implies that there exists severe restrictions on the available functional forms of the conditional probability distributions in order to achieve a consistent joint probability structure. To illustrate the consistency requirement, consider the following 4 x 4 rectangular process in Figure 3.1. There are two arbitrary labelings of sites shown in Figure 3.1. The consistency lemma by Kolmogorov says that any marginal joint cdf's should be the same for both labelings. Therefore, the joint probability function of site variables x_3, x_4, x_7, x_8 should be the same as that of x_9, x_{13}, x_{10}, x_{14} according to the two respective labelings, etc.

1 1	2 5	3 9	4 13
5 2	6 6	7 10	8 14
9 3	10 7	11 11	12 15
13 4	14 8	15 12	16 16

Figure 3.1. A 4 x 4 rectangular lattice process with two labelings of sites.

This problem of consistency has been investigated by Levy (28), Brook (19), Spitzer (33), Hammersley and Clifford (27) and Besag (23,37). We shall describe Hammersley and Clifford and Besag's method of solving this problem.

The second difficulty with equation (3.3) is that the expressions for absolute probabilities involve extremely awkward normalizing function with the consequence that direct approaches to statistical inference through the likelihood function are rarely possible.

3.2.2 Markov Property and Hammersley - Clifford Theorem

In this section, we examine the constraints on the functional form of the conditional probability function at each site. A theorem by Hammersley - Clifford (27) and a simple alternative proof by Besag (37) will be introduced. We begin by describing the problem more precisely.

The first definition determines the set of neighbors for each site (37). Site $j(\neq i)$ is said to be a neighborhood of site i if and only if the functional form of $P(x_i/x_1,...x_{i-1},x_{i+1},...x_n)$ is dependent upon the variable x_j. For the nearest-neighbor system, $P(x_{i,j}/$ all other site values) depends only upon $x_{i,j}$, $x_{i-1,j}$, $x_{i,j-1}$, $x_{i,j+1}$ for each internal site (i,j). Note that this definition of neighbor is different from the one given in deriving compound decision rules. The definition in Chapter 2 defines neighboring cells as any two cells with a common side and contextual information between nonadjacent cells are negligible. From that definition, we used, for example, simplifications like $P(K_3/K_4,\theta_k) = P(K_3/\theta_k)$ (Refer to Figure 2.1). However this is no longer true for the new neighborhood definition given above since the joint probability of K_3, K_4 (Figure 2.1) will indicate that x_3 and x_4 are not independent. The new definition of neighboring cells is also reasonable. Imagine that the neighbors of K_3 include K_8 and the neighbors of K_8 include K_4, there would be some correlation existing between K_3 and K_4 (Figure 2.2). A compound decision rule completely ignores these kind of correlations.

Any set of n sites, each with specified neighbors, generates a class of valid stochastic schemes. Any member of this class is called a Markov field. Any set of sites which either consists of a single site or else in which every site is a neighbor of every other site in the set is called a clique. Thus in the NB system, there are cliques of the form $\{(i,j)\}$, $\{(i-1,j), (i,j)\}$, and $\{(i,j), (i,j-1)\}$ over the entire lattice (37). The definition of clique will be used in constructing valid Markov fields.

We now make two assumptions following Hammersley and Clifford (27). First, we suppose that there are only a finite number of values available at each site. Secondly, we assume that the value zero is available at each site. This second assumption is to ensure that an entire realization of zeros is possible. That is $P(\underline{0})>0$ and one may now define

$$Q(\underline{X}) = \mathrm{Ln}\{P(\underline{X})/P(\underline{0})\} \tag{3.5}$$

for any $\underline{X}\varepsilon\Omega$.

Write $\underline{X}_i$ for the realization $(x_1,...x_{i-1}, 0,x_{i+1},...x_n)$. Now the problem is to find a most general form which $Q(\underline{X})$ may take in order to give a valid probability structure to the system with given neighbors of each site (37). Since

$$\exp\{Q(\underline{X}) - Q(\underline{X}_i)\} = P(\underline{X})/P(\underline{X}_i)$$

$$= P(x_i/x_1,...x_{i-1},x_{i+1},...x_n)/P(0/x_1,...x_{i-1}, x_{i+1},...,x_n)$$

We see that the solution also gives the most general form which may be taken by the conditional probability distribution at each site.

Hammersley-Clifford theorem (27) enables us to write down the most general form of $Q(\underline{X})$ and hence for the conditional distributions. However the original proof is circuitous and requires the development of an operational calculus (the blackening algebra). A simple and interesting proof by Besag (37) will be given instead. (It can be skipped on first reading.)

First note that for any probability distribution $P(\underline{X})$, there exists an expansion of $Q(\underline{X})$, unique on Ω and of the form

$$Q(\underline{X}) = \sum_{1\le i\le n} x_i G_i(x_i) + \sum\sum_{1\le i<j\le n} x_i x_j G_{ij}(x_i,x_j)$$

$$+ \sum\sum\sum_{1\le i<j<k\le n} x_i x_j x_k G_{i,j,k}(x_i,x_j,x_k)+....$$

$$+x_1 x_2...x_n G_{1,2...,n}(x_1,...,x_n) \tag{3.7}$$

For example, we have

$x_i G_i(x_i) = Q(0,\ldots 0,x_i,0,\ldots 0)-Q(\underline{0})$

and also $x_i x_j G_{ij}(x_i,x_j) = Q(0,\ldots x_i,\ldots x_j,\ldots 0)$ $-Q(0,\ldots x_i \ldots 0)-Q(0, \ldots x_j, \ldots 0)+2Q(\underline{0})$. With the above notation, the Hammersley and Clifford's theorem may be stated as follows:

<u>Theorem 3.1</u>:

For any $1\leq i<j\ldots<s\leq n$, the function $G_{i,j,---,s}$ in (3.7) may be nonnull if and only if the sites $i,j,\ldots,s$ form a clique. Subject to this restriction, the G-functions may be chosen arbitrarily.

Thus given the neighbors of each site, one can immediately write down the most general form for $Q(\underline{X})$ and hence for the conditional distributions. We shall see examples of this later on.

<u>Proof</u>:

It is apparent from (3.6) that $Q(\underline{X})-Q(\underline{X}_i)$ can only depend upon x_i itself and the values at sites which are neighbors of site i. Without loss of generality, we shall consider site 1 in detail. We have, from (3.7),

$$Q(\underline{X})-Q(\underline{X}_1) = x_1\{G_1(x_1) + \sum_{2\leq j\leq n} x_j G_{1,j}(x_1,x_j) +$$

$$\sum\sum_{2\leq j<k\leq n} x_j x_k G_{1,j,k}(x_1,x_j,x_k)+\ldots$$

$$+x_2x_3\ldots x_n G_{1,2,\ldots n}(x_1,x_2,\ldots,x_n)\}$$

Now suppose site $\ell(\neq 1)$ is not a neighbor of site 1. Then $Q(\underline{X})-Q(\underline{X}_1)$ must be independent of x_ℓ for all $\underline{X}\varepsilon\Omega$. Putting $x_i=0$ for i=1 or ℓ, we see that $G_{1,\ell}(x_1,x_\ell)=0$ on Ω. Similarly by other suitable choices of $\underline{X}$, it is easily seen that all 3-, 4-,... n-variable G-functions involving both x_1 and x_ℓ must be zero.

The analogous result holds for any pair of sites which are not neighbors of each other and therefore,

$G_{i,j,...,s}$ can only be non-zero if sites $i,j,...s$ form a clique.

On the other hand, any set of G-functions gives rise to a valid probability distribution $P(\underline{X})$ which satisfies the positivity condition. Also since $Q(\underline{X})-Q(\underline{X}_i)$ depends only upon x_ℓ if there is a non-null G-function involving both x_i and x_ℓ, it follows that the same is true for $P(x_i/x_1,...x_{i-1},x_{i+1},...,x_n)$. This completes the proof.

In practice, we usually find that the sites occur in a finite region of Euclidean space and that they often fall into two sets: those which are internal to the system and those which form its boundary. In constructing a Markov field, it is easy for us to make reasonable assumptions concerning the conditional distribution associated with each of the internal sites but problems arise at the boundary of the system. Such problems may be by-passed by considering the joint distribution of the internal site variables conditional upon fixed boundary values. We then need to only specify the neighbors and associated conditional probability structure for each of the internal sites in order to define uniquely the above joint distribution.

Finally in this section, we note that for discrete variables, a further proof of the Hammersley-Clifford theorem has been given by Grimmett (26). Other references on the specification of Markov fields include Averintsev (56), Preston (34), Spitzer (33) and Sherman (35). We are now in a position to see how we can obtain more specific models and schemes.

3.2.3 Auto-models and Gaussian Schemes

We now become more specific in the discussion of spatial schemes (37). Given n sites, labeled 1,.., n, and the set of neighbors for each, we have seen how to construct a general probability distribution associated with the site variables $x_1,..,x_n$. Within this general framework, one is concerned with the particular schemes for which $Q(\underline{X})$ has the representation

$$Q(\underline{X}) = \sum_{1 \leq i \leq n} x_i G_i(x_i) + \sum\sum_{1 \leq i < j \leq n} \beta_{i,j} x_i x_j \qquad (3.8)$$

Where $\beta_{i,j}=0$ unless sites i and j are neighbors of each other. Such schemes will be called Auto-models (37). Note that in the NB case, auto-model will give exact representation for $Q(\underline{X})$ since there is no clique having three or more sites. The probability structure of the system for auto-models is then dependent only upon contributions from cliques containing no more than two sites.

One assumption made in considering auto-models is that the conditional probability distribution associated with each of the sites belongs to the exponential family of distributions (37). That is, for each i,

$$\text{Ln } P(x_i/\cdot) = A_i(\cdot)B_i(x_i) + C_i(x_i) + D_i(\cdot) \qquad (3.9)$$

Where $P(x_i/\cdot) = P(x_i/$ neighbors of site i) and B_i and C_i are of specified form and A_i, D_i are functions of the values at sites neighboring site i. A valid choice of A_i determines the type of dependence upon neighboring site values and D_i is then the appropriate normalizing function.

It can be shown that A_i must satisfy (37)

$$A_i(\cdot) = \alpha_i + \sum_{j=1}^{n} \beta_{i,j} B_j(x_j) \qquad (3.10)$$

Where $\beta_{j,i} = \beta_{j,j}$ and $\beta_{i,j}=0$ unless sites i and j are neighbors of each other. Hence, it follows that G_i in (3.8) has the form

$$G_{i,j}(x_i,x_j) = \beta_{i,j} H_i(x_i) H_j(x_j) \qquad (3.11)$$

Where $x_i H_i(x_i) = B_i(x_i) - B_i(0)$. Thus one generates the class of auto-models by making the additional requirement that, for each i, the function B_i is linear in x_i. The linearity of B_i is satisfied by the most common members of the exponential family.

It is clear that, in terms of equation (3.8), auto-models have conditional probability structure satisfying

$$P_i(x_i/\cdot)/P_i(0/\cdot) = \exp[x_i\{G_i(x_i) + \sum_{j=1}^{n} \beta_{i,j}x_j\}] \quad (3.12)$$

where again $\beta_{j,i}=\beta_{i,j}$ and $\beta_{i,j}=0$ unless site i and j are neighbors of each other (37).

In many practical situations, especially those arising in plant ecology, it is reasonable to assume that the joint distribution of the site variables (plant yields) is multivariate normal. It is evident that any such scheme is an auto-normal scheme. One is particularly interested in the scheme for which

$$P_i(x_i/\cdot) = (2\pi\sigma^2)^{-1/2} \exp[-\frac{1}{2}\sigma^{-2} \{x_i-\mu_i-\Sigma\, \beta_{i,j}(x_j-u_j)\}^2] \quad (3.13)$$

From equation (3.3), one can find the joint density function as

$$P(\underline{X}) = (2\pi\sigma^2)^{-n/2}\, |\underline{B}|^{1/2} \exp\{-\frac{1}{2}\sigma^{-2}(\underline{X}-\underline{U})^T\underline{B}(\underline{X}-\underline{U})\} \quad (3.14)$$

Where $\underline{U}$ is the nx1 vector of finite means, μ_i, and $\underline{B}$ is the nxn matrix whose diagonal elements are unity and whose off-diagonal (i,j) element is $-\beta_{i,j}$ (37). It requires $\underline{B}$ to be positive definite in order for the formulation to be valid.

Now we have presented a rectangular lattice process with given neighbors of each site. We will see, in next section, how one can use this rectangular lattice process to implement a contextual classification algorithm.

3.2.4 Application of Rectangular Lattice Process

To make the best use of rectangular lattice process, the development of a contextual decision rule is of utmost importance. For the purpose of this study, we shall concentrate our attention only on the nearest neighbor system. It is not believed that much information can be gained by considering second or higher order neighbor systems. Besides, the computational complexity will be so significant for the second or higher order neighbor systems that this information will not be used to advantage. Then, given the patterns x_k, x_{k_1}, x_{k_2}, x_{k_3}, x_{k_4} (Figure 2.1), the problem is to decide to which class cell k should be classified. If an optimum Bayesian strategy is to be employed, the decision rule is to choose that class which maximizes the posteriori probability $P(\theta_k/x_k, x_{k_1}, x_{k_2}, x_{k_3}, x_{k_4})$. Equivalently we can maximize the quantity $P(x_k, x_{k_1}, x_{k_2}, x_{k_3}, x_{k_4}/\theta_k)\cdot P(\theta_k)$, where $P(\theta_k)$ is the prior probability of occurrence of class θ_k. Clearly, in order to evaluate $P(x_k, x_{k_1}, x_{k_2}, x_{k_3}, x_{k_4}/\theta_k)$, the classes of cells at sites k_1, k_2, k_3, k_4 must be known. However, the neighboring classes are unknown to us. This implies that either these neighboring true classes have to be estimated using some other preclassification algorithm such as simple decision rule, or they can be classified simultaneously with the cell being considered (cell k). In the latter case, rather than having an unknown class θ_k, we now have an unknown "class-vector" $\underline{\theta}_k = (\theta_k, \theta_{k_1}, \theta_{k_2}, \theta_{k_3}, \theta_{k_4})$. In the present study, we would like to investigate the class-vector case.

Tabulation of the frequency, $P(\underline{\theta}_k)$, for typical imagery frames of interest becomes a challenging problem because of the growth of memory requirement due to large contextual neighborhood (window) size and large number of classes. Remember that the contextual decision rule is now to choose that class vector which maximizes $P(x_k, x_{k_1}, x_{k_2}, x_{k_3}, x_{k_4}/\underline{\theta}_k)\cdot P(\underline{\theta}_k)$.

Consider the pixel array in Figure 3.2 and the simple case where the context window size is three and the number of classes is two as shown in Figure 3.3.

A	F	A	A
F	A	F	F
F	A	A	F
F	A	A	F

A and F represent distinct classes

Figure 3.2 Sample pixel array for frequency distribution determination

Window size = 3
Number of classes = 2

x_3 x_1

x_2

Possible Configurations

x_1	x_2	x_3
A	A	A
A	A	F
A	F	A
A	F	F
F	A	A
F	A	F
F	F	A
F	F	F

Figure 3.3 Sample context problem

To tabulate the frequency distribution, the specified window is placed over the pixel array beginning in the upper left corner. A count associated with this combination of classes inside the window is incremented. The window is moved one pixel to the right, and the process repeated until the specimen pixel array has been exhausted. The simplest solution is to reserve memory for each possible combination and to store the associated frequency. Consider the fairly typical situation, however, in which the window size is nine (three by three neighborhood) and the number of classes is seventeen. Each three by three block can take one of 17 to the ninth power (or 120,000,000,000) possible configurations and this simple approach is obviously infeasible.

Note that a given scene, having finite size, must contain only a small fraction of this huge number. For example, consider classifying a 200 pixel by 200 pixel scene. In the worst case there are 40,000 different combinations present. Many of these 40,000 will be repeated often, further reducing the number of unique combinations. Only combinations which actually occur need to be stored.

Figure 3.4 summarizes the presently implemented approach. Each time a combination is obtained, the binary tree is traversed until the combination is found or until a null node is encountered. If the combination is already in the tree, the count associated with that combination is incremented. Otherwise this is the first time this combination has occurred, and therefore a new node is added to the tree and its associated count is initialized to one.

For our frequency tabulation, the frequency distribution was determined from a classification results obtained by a professional analyst. The study area is again taken from Tippecanoe County, Indiana. The pixel array used to tabulate frequency distribution for the nearest-neighbor window is of size 128 x 128 (lines 125-252, columns 187-314). The frequency distribution for the first seventy-four class-vectors is shown in Table 3.1, where the number of classes is seventeen.

The ratio of the actual number of unique combinations to the possible number of unique combinations (a function of scene size) provides an indication of the "amount" of context in the scene. The scene-context ratio is always greater than zero and less than or equal to one. For a scene with frequently occurring combinations, the numerator will tend to

1. Input — Classification results file
window configuration
image size
2. Get Combination
Combination found
a. increment count
3. Tree — Null node
a. create node
b. count = 1
4. If more combinations Go to 2
5. Print Tree using in order traversal

Figure 3.4 Binary Tree Implementation.

Class Vectors	Prior Probabilities	Class Vectors	Prior Probabilities
1 1 1 1 1	0.000111	1010101010	0.013227
2 2 2 2 2	0.109595	1111111111	0.159358
2 2 2 2 3	0.023305	1111111112	0.029604
2 2 2 2 4	0.019526	1111111211	0.018266
2 2 2 3 2	0.014487	1111121111	0.050390
2 2 2 4 2	0.017636	1111121211	0.020156
2 2 3 2 2	0.023935	1112111111	0.017006
2 2 4 2 2	0.023935	1112111112	0.013857
2 3 2 2 2	0.017635	1112121111	0.014487
2 4 2 2 2	0.013857	1211121211	0.015747
3 2 3 3 3	0.017635	1211121212	0.015117
3 3 2 2 3	0.013227	1212111212	0.017636
3 3 2 3 3	0.030234	1212121111	0.013227
3 3 3 3 2	0.023935	1212121112	0.013857
3 3 3 3 3	0.376664	1212121211	0.031494
3 3 3 3 5	0.040942	1212121212	0.057948
3 3 3 3 6	0.015747	1313131313	0.034643
3 3 3 5 3	0.032124	1414141414	0.147390
3 3 3 5 5	0.013857	15 7151515	0.013857
3 3 4 3 3	0.018266	1515 71515	0.017006
3 3 5 3 3	0.033383	1515141515	0.020156
3 3 5 5 3	0.017006	15151515 7	0.015117
3 5 3 3 3	0.033383	1515151514	0.015117
3 5 3 3 5	0.020156	1515151515	0.120306
4 4 4 4 4	0.025826	16 5 51616	0.020786
5 3 5 5 5	0.015117	16 51616 5	0.028344
5 4 5 5 5	0.018266	16 5161616	0.028974
5 5 3 5 5	0.018896	1616 5 516	0.029604
5 5 4 5 5	0.017636	1616 516 5	0.013857
5 5 5 3 5	0.017000	1616 51616	0.047980
5 5 5 5 3	0.018896	161616 5 5	0.015117
5 5 5 5 5	0.177624	161616 616	0.028974
6 6 6 6 6	0.061000	16161616 5	0.058578
7 7 7 7 7	0.061098	1616161616	0.197150
7 7 7 7 18	0.015117	1717171717	0.015747
8 8 8 8 8	0.078734		
8 8 9 8 8	0.016376		
9 9 9 9 9	0.073065		
9 9 9 9 12	0.013227		

Table 3.1. A priori probabilities of 74 class-vectors.

be small, resulting in a low scene context ratio and indicating a high-context scene.

The next step in developing a contextual decision rule is to evaluate the class-vector conditional joint probability density function $P(x_k, x_{k_1}, x_{k_2}, x_{k_3}, x_{k_4}/\underline{\theta}_k)$. Suppose that a lattice process (3.14) is defined for all site variables on a fixed-size rectangular lattice. Then, the variances and co-variances for the site variables within the nearest neighbor (NB) window must be extracted from the huge-dimensioned variance-covariance matrix $\sigma^2\underline{B}^{-1}$ for all the lattice points. This creates the following problem. Since the lattice process defined in (3.14) is not stationary (readers can verify this by constructing $\underline{B}$ matrix for a small pixel array and finding its inverse), the variance-covariance matrix for the points in the NB window has to be calculated from some analytic expression every time the window is moved in the rectangular lattice to classify the next pixel. Although the finding of such an explicit expression for the variances and covariances of the lattice points is possible, the extraneous calculations for the variance-covariance matrices every time a new point in the lattice is to be classified make the rectangular lattice process impractical.

Naturally, a stationary lattice process is more desirable because correlations between site variables are not position-dependent. However, because of the rectangular lattice boundary effect, fixed sized lattice process developed in the last section will not be stationary. Nevertheless, we can make some approximations to reach our goal.

Remember in the construction of the spatial processes that the set of spatial points is arbitrary as long as there are spatial interactions existing among them. Imagine that a process is defined on the five points in the NB system! From the configuration and labeling given in Figure 3.5, the $\underline{B}$ matrix for the five site variables in the NB system is

	x_3	
x_4	x_1	x_2
	x_5	

Figure 3.5 Labeling of NB points

$$\underline{B} = \begin{bmatrix} 1 & -\beta_{12} & -\beta_{13} & -\beta_{14} & -\beta_{15} \\ -\beta_{21} & 1 & 0 & 0 & 0 \\ -\beta_{31} & 0 & 1 & 0 & 0 \\ -\beta_{41} & 0 & 0 & 1 & 0 \\ -\beta_{51} & 0 & 0 & 0 & 1 \end{bmatrix}$$

Since the LANDSAT data is symmetric, it is reasonable to assume that the system is both homogeneous and isotropic, i.e. all four β's in $\underline{B}$ are equal. In this case we find that the variance of cell 1 is proportional to $1/(1-4\beta^2)$ and that of the other cells is proportional to $(1+\beta^2)/(1-4\beta^2)$. Since $\beta<0.5$, the variances of these five cells are approximately equal if we assume $(1+\beta^2)\simeq 1$. The covariances between each pair of the five points are found to be stationary; that is, the covariances depend only on the distance between two points but not on their location in the lattice. In this way, we have approximated a stationary process where the correlations can be calculated once for all without repeating every time we move the window. All five cell variables in the window placed over the pixel array are assumed to follow this model.

In real situations, many different ground cover types are present in LANDSAT data. These classes are characterized by different covariances and thus the model will not fit the data directly. However, the classes with large variances can be decomposed into subclasses resulting in smaller variances by clustering. An attempt was made to reduce the class variances by introducing subclasses such that all the class variances were approximately the same. Therefore, classes are now characterized by their means only. Even though this is an awkward and impractical requirement, it allows us to use the rectangular lattice model to introduce context into our decision process. The context employed in the present class-vector approach is the true class directional transition probability context. After investigating the class-vector approach, we will examine the problems with this approach and refine our procedures toward a better utilization of context.

The data block of size 128 x 128 (lines 125 - 252, columns 187-314) was used as training data to estimate the β and σ^2. A maximum likelihood technique is employed to perform the estimation by selecting independent NB samples from the data block (Figure 3.6). Four nonlinear equations arise in the maximization process. Newton-Raphson technique (Appendix C) was used to solve the nonlinear equations. The estimated values of the four β's and the conditional variance σ^2 are given in Table 3.2

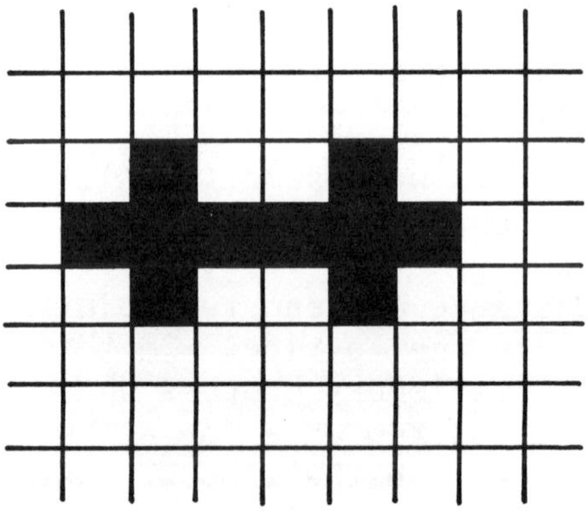

Figure 3.6. Two independent NB samples.

With the estimates of β and σ^2, and the frequency distribution tabulation, it is possible to evaluate the contextual decision rule. When a point is to be classified, a NB window is placed over that pixel, the discriminant function is calculated for each class-vector to make a decision. The one-band (0.7-0.8uM) contextual classification result for the area (line 200-327, columns 120-247) of Lafayette data is shown in Figure 3.7(c). The two-band and one-band simple classification results, used for comparison purposes, are also shown in Figure 3.7(a), (b).

So far we have only considered the univariate site variable case in the model. To consider multivariate site variable problem, the stationary plane lattice process must first be constructed. To this end, we adopted a rather heuristic extension of the univariate site variable case to implement the bivariate site-variable case for our problem. In the extension, the correlations between the two spectral bands are ignored and the resulting $\underline{B}$ matrix is now

$$\begin{matrix}
1 & -\beta_1 & -\beta_2 & -\beta_3 & -\beta_4 & 0 & 0 & 0 & 0 & 0 \\
-\beta_1 & 1 & 0 & 0 & 0 & 0 & 0 & 0 & 0 & 0 \\
-\beta_2 & 0 & 1 & 0 & 0 & 0 & 0 & 0 & 0 & 0 \\
-\beta_3 & 0 & 0 & 1 & 0 & 0 & 0 & 0 & 0 & 0 \\
-\beta_4 & 0 & 0 & 0 & 1 & 0 & 0 & 0 & 0 & 0 \\
0 & 0 & 0 & 0 & 0 & 1 & -\beta_1 & -\beta_2 & -\beta_3 & -\beta_4 \\
0 & 0 & 0 & 0 & 0 & -\beta_1 & 1 & 0 & 0 & 0 \\
0 & 0 & 0 & 0 & 0 & -\beta_2 & 0 & 1 & 0 & 0 \\
0 & 0 & 0 & 0 & 0 & -\beta_3 & 0 & 0 & 1 & 0 \\
0 & 0 & 0 & 0 & 0 & -\beta_4 & 0 & 0 & 0 & 1
\end{matrix}$$

This definitely is not a theoretically-sound way to treat the multivariate site-variable problem. Nevertheless, the classification algorithm is performed on the same test block area to compare the results. The two-band contextual result together with computation times used by IBM 360 computer are

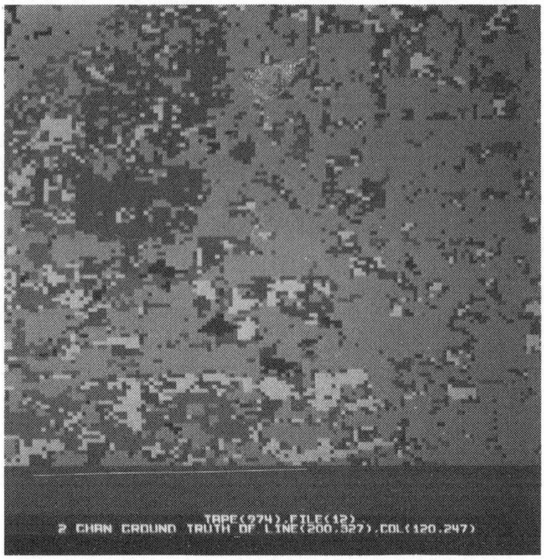

a) 2 band simple rule result (2:34 Min. cpu time used)

b) 1 band simple rule result

Figure 3.7. Class-vector approach results of area (Line 200-327, Columns 120-247).

c) 1-band 4-neighbor rule result (14:7 Min. cpu time used)

d) 2-band 4-neighbor rule result (48:00 Min. cpu time used)

Figure 3.7. (Continued)

Table 3.2 Estimation Results for Block (Lines 125-252, Columns 187-314)

β_1	β_2	β_3	β_4	σ^2
0.43425	0.41961	0.43154	0.40206	4.7871

shown in Figure 3.7 (d). The improvement in performance is discernable in the upper left portion of the picture where large residential areas are present.

3.2.5 Problems and Disadvantages of the Present Approach

A list of the problems encountered in the contextual classification algorithm developed in this section is given below:

A) The approach is not theoretically sound. We have seen that many approximations and assumptions have to be made to implement the decision rule. The major handicap of this approach is that the process defined on a finite rectangular lattice is not stationary. An infinite stationary lattice process will be developed in the next section.

B) The approach is time consuming. In the classification scheme, we are classifying five cells (NB system) simultaneously. This means that we have a lot more class-vectors than the original number of classes to classify. The computer time for classification will undoubtly increase considerably. Our goal in deriving a contextual algorithm is to reduce computation time as much as we can. Also there is no rule to decide how many class-vectors have to be considered.

C) The approach is not reliable. Since we classify all five cells at the same time, a natural question to ask is whether the classification of the neighboring cells (K_1, K_2, K_3, K_4 in the NB system) has any effect on the classification result of the center cell (K in NB system). The answer is that they influence the result to a great extent. We will use Table 3.1 to explain the effect. Suppose we have a case where the center cell is class 2, the four neighboring cells have classes 2, 2, 3, 4 (remember the ordering of neighbors in Figure 3.5). When we classify this NB sample (Figure 3.8), we would naturally have the highest probability for the class-vector 2 2 2 3 4 in Table 3.1. However, this class-vector is not included in the list of Table 3.1. The reason is because the probability of occurrence of this class-vector is below the threshold we selected. What will happen is that, very probably, some other class vector will result in a high probability and the center cell (with true class being 2) is erroneously classified using context. The ideal situation is to include all available class-vectors. But there are too many such class-vectors, as discussed earlier, and the computation time will even be increased much more. Therefore, this is not a desirable approach to use context. Since Table 3.1 was obtained from a set of training data, the testing data would have many different NB class-vectors which are not in Table 3.1. This would result in certain errors before we even start the classification.

	2	
3	2	2
	4	

Figure 3.8 A class-vector configuration

D) The approach is not flexible. The extension into multivariate site variable case is difficult to carry out.

After we have seen these problems with the present approach, let us briefly describe what we intend to do. The first problem to be solved is to define a stationary process on an infinite plane. The extension from rectangular process to infinite lattice process will be discussed in detail in the next section. We shall then consider the multivariate site variable situation and provide some theoretical justification. The final contextual classification algorithm will be proposed and experimental results will be given to show the efficiency of the developed algorithm in Chapter 4.

3.3 Gaussian Markovian Process on Infinite Lattice

Some definitions of Markovian property for a lattice process will be given in this section. A torus lattice is constructed from methods given in the last section. An infinite plane process is then constructed from the torus process. Estimation techniques are also introduced.

3.3.1 Markov Process on a Torus

Moran (40) constructed a stationary process on the square lattice formed by all the pairs of positive and negative integers, in which the random variables are identically normally distributed and satisfy a Markovian property. This is done by first constructing a similar process on a torus lattice and later letting the size of the torus tend to infinity. He considers a torus because the torus process is periodic, and the process is defined if one knows its spectral density function. Therefore, one must be able to write down the spectral density function (which contains all the correlations between pairs of sites) of the infinite lattice process.

The approach that will be taken can best be illustrated by the one dimensional case. Consider the time function (one realization of a process) shown in Figure 3.9. The Fourier series representation obtained will be valid in the time interval from t_1 to t_2 for almost any time function. If, however, $x(t)$ is periodic with a period $T = t_2-t_1$, then the representation is valid for all times and not just those within the interval.

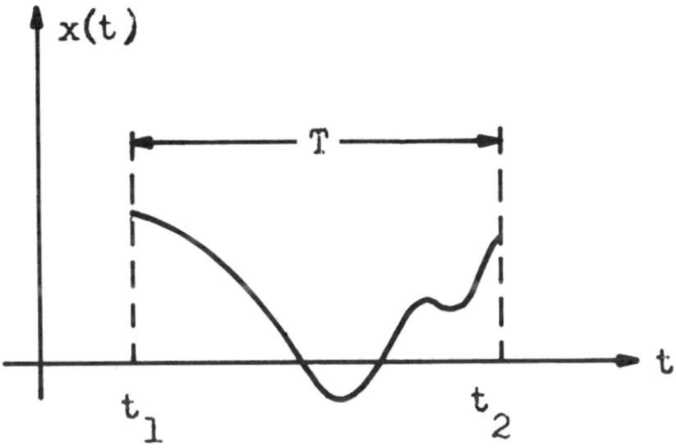

Figure 3.9. Time function to be represented by Fourier series.

If coefficients for the Fourier series are designated as α_n, then they may be expressed as

$$\alpha_n = \frac{1}{T}\int_{t_1}^{t_1+T} x(t)e^{-jnw_o t}\, dt \qquad (3.15)$$

where $w_o = \frac{2\pi}{T}$ and are usually complex. Then x(t) is given by

$$x(t) = \sum_{n=-\infty}^{\infty} \alpha_n e^{jnw_o t} \qquad (3.16)$$

This time function can be precisely represented over the interval $-\frac{T}{2} < t < \frac{T}{2}$ (moving the function along t-axis) by means of a Fourier series having a period of T. As the period T is increased, more and more of the time function will be included in the series representation. In the limit as $T \to \infty$, the entire function x(t) will be included in the expression. The validity of the resulting expression will be determined by the nature of the original time function x(t) and is dependent on whether the mathematical operations are allowable or not. As we said earlier, the extension is unique. Consider now the representation of x(t) as a complex Fourier series over the interval of $-\frac{T}{2} < t < \frac{T}{2}$.

$$x(t) = \sum_{n=-\infty}^{\infty} \alpha_n \exp(j \frac{2\pi nt}{T}) \tag{3.17}$$

where

$$\alpha_n = \frac{1}{T}\int_{\frac{-T}{2}}^{\frac{T}{2}} x(t) \exp(-j \frac{2\pi nt}{T}) \tag{3.18}$$

The fundamental angular frequency is $w_o = 2\pi/T$. Using this expression for w_o and substituting (3.18) into (3.17) gives

$$x(t) = \sum_{n=-\infty}^{\infty} \exp(j\frac{2\pi nt}{T}) \frac{w_o}{2\pi}\int_{\frac{-T}{2}}^{\frac{T}{2}} x(t) \exp(-jnw_o)dt \tag{3.19}$$

If we now let $T\to\infty$, the spacing between harmonics will become a differential, that is, $w_o = 2\pi/T\to dw$; the number of components becomes infinite, that is, $n\to\infty$; the angular frequency of any particular component is given by nw_o; and the summation formally passes into an integral. Equation (3.19) may then be written in the following form:

$$x(t) = \int_{-\infty}^{\infty} e^{jwt} \left[\frac{dw}{2\pi}\int_{-\infty}^{\infty} x(t)\ e^{-jwt}dt\right] \tag{3.20}$$

Rearranging terms gives

$$x(t) = \frac{1}{2\pi}\int_{-\infty}^{\infty} [\int_{-\infty}^{\infty} x(t)\ e^{-jwt}dt]\ e^{jwt}\ dw \tag{3.21}$$

This is the Fourier integral relation. The inner integral is called the Fourier transform of x(t) and is designated as

$$F\{x(t)\} = F(jw) = \int_{-\infty}^{\infty} x(t)\ e^{-jwt}dt \tag{3.22}$$

The relationship in (3.21) is called the inverse Fourier transform of F(jw) and is written as

$$F^{-1}\{F(jw)\} = x(t) = \frac{1}{2\pi}\int_{-\infty}^{\infty} F(jw)\ e^{jwt}\ dw \qquad (3.23)$$

The functions x(t) and F(jw) are called Fourier transform pairs. Here we see that a Fourier series representation is defined for a periodic time function. The Fourier representation is then extended to include the time function defined on the infinite line. This is the way that Fourier transform (frequency spectrum representation of time function) may be obtained for an infinite time function.

Moran takes a similar approach to define an infinite plane lattice process. This process will be stationary because there would be no boundaries and one may say that stationarity is reached for all the covariance functions. The first thing we need to do is to have available for us a lattice process defined on periodic rectangular lattice functions. To assure that the rectangular lattice function (one realization of a rectangular process) is periodic, we can form a torus lattice from a rectangular lattice.

Let the torus be defined by all pairs of integers (j,k), j=0,...p-1, k=0, ... q-1 where p and q are identified with zero. There are then pq random variables X_{jk} at points (j,k) for the univariate-site variable case. We will consider multivariate-site variables situation in a later section. Since the torus lattice is formed from a rectangle lattice without changing the orientations of its sites, the developed theory can be directly applied to torus lattices. We shall restrict our attention to the NB system because it has been found that the computation time will be increased too much to compensate for the accuracy we will gain by using higher order schemes. Meanwhile, we will assume that the system is both spatially homogeneous and isotropic. That is, the G functions in (3.7) are independent of positions. Equivalently the β's in $\underline{B}$ are all the same. The validity of this assumption will be given later

The joint normal density function (3.14) will then

be

$$p=K\ \exp[-\tfrac{1}{2}\{\Sigma x_{jk}^2+a\Sigma x_{jk}(x_{j+1,k}+x_{j-1,k}+x_{j,k+1}+x_{j,k-1})\}]$$

(3.24)

where $a = -\beta$ and $|a|<\frac{1}{4}$ and K is a constant (40). This is a proper statistical distribution because it can be shown that the quadratic form inside the brackets is positive definite. To see this, we observe that any cross product is similar to $x_{jk}x_{j+1,k}$ and occur twice in the sum, and that

$$|2a\ x_{jk}x_{j+1,k}|<\frac{1}{4}(x_{jk}^2+x_{j+1,k}^2) . \qquad (3.25)$$

We state some definitions of Markovian property and show that this torus process possesses a Markovian property.

Definition 3.1. $\{X_{st}\}$ is weakly Markovian if the following condition is satisfied. Let A be any finite set of points on the lattice, and X_A the corresponding set of variates. Let ∂A be the boundary of A, i.e., the set of points of the lattice not in A which are joined by A by a single bond. Let $X_{\partial A}$ be the set of variates on this boundary. Let B be any finite set of points not in A or ∂A, and X_B the corresponding set of variates. Then we say that $\{X_{st}\}$ is weakly Markovian if the resulting probability distributions satisfy the condition

$$P(X_A,X_B/X_{\partial A}) = P(X_A/X_{\partial A})P(X_B/X_{\partial A}) \qquad (3.26)$$

This is the definition used by Hammersley and Clifford (27).

Definition 3.2. The process $\{X_{st}\}$ is strongly Markovian if the following condition is satisfied. Let ℓ be any closed noninersecting line on the lattice, i.e., consisting of a sequence of joints between neighboring points which form a closed finite non-self-intersecting line. Let X_A be the set of the variates corresponding to points interior to ℓ, X_ℓ the set of variates corresponding to points on ℓ, and X_B any set of points outside ℓ. Then we must show that for all X_ℓ such that $P(X_\ell)>0$

$$P(X_A,X_B/X_\ell) = P(X_A/X_\ell)P(X_B/X_\ell) \tag{3.27}$$

For the following definition we assume that X_{st} have a finite variance. Now define

$$\rho_{st} = \text{Correlation } (X_{\infty},X_{st}) = \rho_{-s,t}=\rho_{s,-t}=\rho_{-s,-t}$$

<u>Definition 3.3</u>. $\{X_{st}\}$ is weakly Markovian in the sense of Doob (114), if in Definition 3.1 the equation (3.26) is replaced by the condition that all the partial correlations of X_A with X_B, with respect to the values of X_{A}, are zero.

<u>Definition 3.4</u>. $\{X_{st}\}$ is strongly Markovian in the sense of Doob if in Definition 3.2 the condition (3.27) is similarly replaced by the condition that the corresponding partial correlations are zero, with respect to X_ℓ.

<u>Theorem 3.2</u>: Whether the variance exists or not, Definition 3.1 implies Definition 3.2 (41).

Proof: Let ℓ be any closed non-intersecting line on the lattice. Let X_A be the variates corresponding to points interior to ℓ, and X_B variates corresponding to any finite set of points outside ℓ. Let $X_{\partial A}$ be the variates corresponding to points which are neighboring points of points in A (i.e. joined by a single bond), and X_C the variates corresponding to points on ℓ which are not on ∂A. Then if Definition 3.1 is satisfied one knows that

$$P(X_A, X_B, X_C / X_{\partial A}) = P(X_A / X_{\partial A}) P(X_B, X_C / X_{\partial A}) \qquad (3.28)$$

Observing that $P(X_A / X_{\partial A}) = P(X_A / X_{\partial A}, X_C)$ by Definition 3.1, and dividing by $P(X_C / X\partial A)$, which is not zero, because $P(X_\ell) > 0$, one has

$$\frac{P(X_A, X_B, X_C / X_{\partial A})}{P(X_C / X_{\partial A})} = P(X_A / X_A, X_C) \frac{P(X_B, X_C / X_{\partial A})}{P(X_C / X_{\partial A})},$$

so that

$$P(X_A, X_B / X_{\partial A}, X_C) = P(X_A / X_{\partial A}, X_C) P(X_B / X_{\partial A}, X_C) \quad (3.29)$$

which proves the result.

Moran shows that the torus lattice process possesses the natural generalization of the Markovian property.

Let L be any finite set of n points on the lattice and (changing the notation) let $X_1, \ldots X_{n_1}$ be the corresponding set of random variables. Write ∂L for the boundary of the set L, i.e., the set of all lattice points not in L which are joined to L by a unit line on the lattice. Let $X_1', \ldots X_{n_2}'$ be the random variables associated with the points of ∂L. Let M be any finite set of points outside $L+\partial L$, and $X_1'', \ldots, X_{n_3}''$ be the corresponding set of random variables. Then one says that the process satisfies the Markovian property if the distribution of $(X_1, \ldots, X_{n_1})$ conditional on $(X_1', \ldots X_{n_2}')$ is independent from the distribution of $(X_1'', \ldots X_{n_3}'')$ conditional on the same values of $(X_1', \ldots, X_{n_2}')$. Since the distribution (3.24) is multivariate normal this is a property depending only on the correlation matrix of the whole set of variables.

From (3.24) we see that, because (3.24) can be written in the form

$$g_1(X_i, X_j')g_2(X_j')g_3(X_j', X_k'') \tag{3.30}$$

the joint probability density of $(X_1, \ldots, X_{n_1})$, is dependent only on their values and the values of $(X_1', \ldots, X_{n_2}')$, and similarly for $(X_1'', \ldots, X_{n_3}'')$. Thus the process is certainly Markovian.

Moran also defines the Markovian property in another way. Let ℓ be a closed non-intersecting line on the lattice (i.e., consisting of segments joining neighboring points) with an interior and an exterior. Define L to be all sets of points interior of ℓ, replace ∂L by ℓ, and M by any finite set of points outside ℓ. ℓ will include ∂L and may have other points as well. Then the above argument goes through in exactly the same way and the distribution of the variates corresponding to L, conditional on the values of those on ℓ, will be independent from the distribution of the variates corresponding to M conditional on the same values of the variates on ℓ.

Next Moran finds the variance-covariance matrix of the distribution of the X's by inverting the matrix of the coefficients of the quadratic form in (3.24) or equivalently, the $\underline{B}$ matrix in (3.14). After finding the variance-covariances for the torus lattice variables, he was able to find those for the infinite lattice process by extending p, q to infinity. The pq random variables are written as a vector $\underline{X}$ equal to $(X_{00}, X_{01}, \ldots, X_{0,q-1}, X_{11}, \ldots, X_{1,q-1}, \ldots, X_{p-1,q-1})$. The matrix $\underline{B}$ in (3.14) is expressed as $(C_{u_1-u_2, v_1-v_2})$ where $C_{u_1-u_2, v_1-v_2}$ is the element in the row u_1q+v_1 and column u_2q+v_2 where $u_j=0,1,\ldots p-1$ and $v_j=0,1,\ldots q-1$

and their sums and differences are taken modulo (p) and (q) respectively. The matrix $\underline{B}$ is constructed as follows. Consider the pxp circulant matrix

$$\begin{bmatrix} x & y & o & \cdot & \cdot & \cdot & \cdot & y \\ y & x & y & \cdot & \cdot & \cdot & \cdot & 0 \\ \cdot & \cdot & \cdot & \cdot & \cdot & \cdot & \cdot & \cdot \\ \cdot & \cdot & \cdot & \cdot & \cdot & \cdot & \cdot & \cdot \\ y & o & o & \cdot & \cdot & \cdot & y & x \end{bmatrix} \tag{3.31}$$

and replace the x, y, and zeros by $\underline{A}$, $\underline{Q}$, and zero qxq matrices respectively where $\underline{A}$ is the pxp matrix

$$\begin{bmatrix} 1 & a & o & \cdots\cdots & o \\ a & 1 & a & o\cdots\cdots & o \\ o & a & 1 & a\cdots\cdots & o \\ \cdots & \cdots & \cdots & \cdots & \cdots \\ a & o\cdots\cdots & o & a & 1 \end{bmatrix} \tag{3.32}$$

and $\underline{Q}$ is a pxp diagonal matrix with a's down the main diagonal. This represents $\underline{B}$ as a matrix depending on pq quantities which in fact only take the three values 1, a, 0, and

$$\begin{aligned} c_{k\ell} &= 1, \quad \text{if } k = 0,\ \ell = 0, \\ &= a, \quad \text{if } (k=0,\ \ell=1),\ (k=0,\ \ell=q-1),\ (k=1,\ell=0), \text{ or} \\ &\qquad (k = p-1,\ \ell=0), \\ &= 0, \text{ otherwise} \end{aligned} \tag{3.33}$$

Moran then shows that $\underline{B}$ can be diagonalized by a unitary matrix U so that U*BU is a diagonal matrix. For $u_i=0,1,\ldots, p-1$, $v_j=0,1,\ldots, q-1$ he defines the element in the row u_1q+v_1 and column u_2q+v_2 to be

$$(pq)^{-1/2}\exp i\{2\pi u_1 u_2^{p-1} + 2\pi v_1 v_2^{q-1}\} \tag{3.34}$$

It is easy to verify that $U^*U=I$.

The matrix U^*BU equals Λ, a diagonal pq x pq matrix with diagonal elements λ_{tt}, where $t=u_1q+v_1$, given by

$$\lambda_{tt} = \sum_{k,\ell} C_{k\ell} \exp 2\pi i\{k\ u_1p^{-1}+\ell v_1q^{-1}\} \qquad (3.35).$$

In fact $\underline{B}$ contains in each row and each column a single element unity, four elements equal to a, and the rest zeros. On substituting the values (3.33) one gets

$$\lambda_{tt} = 1+a\{\exp(2\pi iq_1q^{-1})+\exp(-2\pi iu_1q^{-1})+\exp(2\pi iv_1p^{-1}) + \exp(-2\pi iv_1p^{-1})\}$$

$$= 1 +2a(\cos 2\pi u_1q^{-1}+\cos 2\pi v_1p^{-1}) \qquad (3.36)$$

Writing $\underline{B}^{-1} = (b_{st})$ where $s = u_1q+v_1, t=u_2p+v_2$. Then

$$B^{-1} = U\Lambda^{-1}U^* \qquad (3.37)$$

and writing $m = u_eq+v_3$ we have

$$b_{st} = \sum_m u_{sm}\lambda_{mm}^{-1}\bar{u}_{mt} \qquad (3.38)$$

$$= \sum_m (pq)^{-1}\exp 2\pi i\{u_1u_3p^{-1}+v_1v_3q^{-1}\}$$

$$\cdot \{1+2a\cos 2\pi v_3q^{-1}+2a \cos 2\pi u_3q^{-1}\}^{-1}$$

$$\cdot \exp - 2\pi i\{u_2u_3p^{-1}+v_2v_3q^{-1}\} \qquad (3.39)$$

This is the real variance-covariance matrix of the random variables on the torus. Notice that (3.39) is a function only of the differences u_1-u_2 and v_1-v_2 and is a symmetric function of each. Thus the process is invariant on the torus under rotations of angles $2\pi p^{-1}$ and $2\pi q^{-1}$, and symmetric under "reflection".

3.3.2 A Markov Process on the Infinite Plane Lattice

Our goal is to construct a stationary Markovian Gaussian process on the (infinite) square lattice formed by all positive and negative integers (m,n). It is well known that it is sufficient to define such a process for all finite sets of such points provided the resulting distribution is invariant under translation. It is therefore sufficient to consider any fixed finite set of points on the torus lattice and then let p an q tend to infinity. Writing $\exp 2\pi i u_3 p^{-1} = z_1$ and $\exp 2\pi i u_3 q^{-1} = z_2$ and letting p and q tend to infinity, Moran finds that the covariance between $X_{m,n}$ and $X_{m+s,n+t}$ is

$$-\frac{1}{4\pi^2} \oint_{|z_1|=1} \oint_{|z_2|=1} \frac{z_1^{s-1}\ z_2^{t-1}}{[1+a(z_1+z_1^{-1})+a(z_2+z_2^{-1})]}\, dz_1\, dz_2 \tag{3.40}$$

the convergence being uniform on the finite set of points (40).

Thus the covariance $V_{s,t}$ between $X_{m,n}$ and $X_{m+s,n+t}$ is generated by the generating function

$$\sum_{-\infty}^{\infty} \sum_{-\infty}^{\infty} z_1^s\, z_2^t\, V_{s,t} = [1+a(z_1+z_1^{-1})+a(z_2+z_2^{-1})]^{-1} \tag{3.41}$$

and we see that the process is not only stationary but also reflection invariant. We define the correlations by $r_{st} = V_{st}V_{oo}^{-1}$. Furthermore, the process is Markovian, for if we take any finite set of points L, ∂L, and M, their joint distribution is the limit of the distributions on the torus and their dependence (since the variates are Gaussian) is dependent only on their variance-covariance matrix which, being the limit of the one for the torus, is such that the distributions of the variates in L and M, conditional on the values of ∂L, are independent.

Thus the process is Markovian.

We can now use the stationary process as a model for our contextual algorithm. The assumption we will have to make is that the data are a partial realization of the infinite lattice process. However, we can also explicitly generate the process on the plane lattice in a direct way. Let E_{st} be independent Gaussian variates with zero means and unit variance. Define the X_{mn} by the series

$$X_{mn} = \Sigma\ \alpha_{m+s,n+t} E_{st} \tag{3.42}$$

where α_{st} is the coefficient of $z_1^s z_2^t$ in the expansion of

$$[1+a(z_1+z_1^{-1})+a(z_2+z_2^{-1})]^{-1/2} \tag{3.43}$$

Because $|a|<\frac{1}{4}$ this expansion is convergent as a Laurent series in a ring $1-z<z_1,\ z_2<1+z$ with $z>0$ (40).

Here we mention Levy's work to construct a process. Levy (51) also constructed a Gaussian process on a lattice which is more general than the one considered here (by Moran (40)) in that he includes in the quadratic form (3.24) also terms of the form $X_{jk}X_{j-1,k-1}$ etc. This is Markovian in the second sense but not in the first. Moreover he does not show strictly that the process exists in that he starts from a formula like (3.24) with an infinite number of lattice points which does not therefore define a distribution, and furthermore does not observe that if we consider the joint distribution of the X_{jk} over a finite set of lattice points, such as on a rectangle, (3.24) does not give the correct distribution since, for example, the X_{jk} would no longer all have the same variance. It is for this reason that we first define the torus lattice, for which (3.24) holds, and then proceed to the limit obtaining the explicit representation (3.44). The correct quadratic form to use for a finite lattice in Levy's case would have to be obtained by the awkward inversion of the matrix of covariances defined by a generalization of (3.44).

One can obtain the covariances explicitly from the integrals

$$V_{st} = \frac{1}{4\pi^2}\int_0^{2\pi}\int_0^{2\pi} \frac{\cos s\theta_1 \cos t\theta_2}{[1+2a\cos\theta_1 + 2a\cos\theta_2]}\, d\theta_1\, d\theta_2 \tag{3.44}$$

Notice that the process defined above is quite different from the nearest neighbor process defined by

$$X_{st} - a(X_{s+1,t}+X_{s-1,t}+X_{s,t+1}+X_{s,t-1}) = E_{st} \tag{3.45}$$

discussed by Whittle (42). This is not Markovian.

Now we consider the evaluation of (3.44) to obtain all the variances and covariances. V_{00} and $V_{10}=V_{-10}=V_{01}=V_{0-1}$ are however simpler. V_{00} has been obtained by Moran (40) explicitly as

$$V_{00} = 2\pi^{-1}\int_0^{1/2\pi} (1-16a^2\sin^2\theta)^{-1/2}d\theta \tag{3.46}$$

which is $2\pi^{-1}$ times the complete elliptical integral of the first kind, tabulated in Alramowitz and Stegun (49).

To obtain the value of V_{10}, we use (3.41) which gives

$$[1+a(z_1+z_1^{-1}+z_2+z_2^{-1})]\Sigma\Sigma z_1^s z_2^t V_{st}=1 \tag{3.47}$$

or

$$\sum_{-\infty}^{\infty}\sum_{-\infty}^{\infty} z_1^s z_2^t V_{st} + a\sum_{-\infty}^{\infty}\sum_{-\infty}^{\infty}(z_1^{s+1}z_2^t + z_1^{s-1}z_2^t + z_1^s z_2^{t+1} + z_1^s z_2^{t-1})V_{st}=1$$

Since the right hand side is a constant, we collect all the constants on the left hand side. This gives

$$V_{00}+a(V_{-10}+V_{10}+V_{0-1}+V_{01})=1$$

so that

$$V_{10}=V_{-10}=V_{01}=V_{0-1}=\frac{1}{4}a^{-1}(1-V_{00})$$

and

$$r_{10}=\frac{1}{4}a^{-1}(V_{00}^{-1}-1) \tag{3.48}$$

The evaluation of integral (3.44), which is known as the lattice Green's function, for other values of m and n it is much more complicated. We use approximations to obtain other values of V_{mn}'s for our experiment later. This integral occurs in the theory of random walks on a lattice, in a number of applications in physics, and has a large literature for which readers may consult Montroll (57), and Katsura and Inawashiro (50).

We have now presented a stationary plane Gaussian Markovian process. Next, we shall show how to estimate the unknown parameters before we finally present the classification algorithm and results.

3.3.3 Estimation of Spatial Interaction in Gaussian Lattice Processes

We shall briefly summarize the spectral density functions for the torus process and the plane lattice process. Estimation equations and specific ones for a first-order scheme (NB system) will be given. A simple coding technique which is used later in experiment is also described.

Let $\{r_{st}:s,t=0,1,...\}$ denote the auto covariances of a stationary Gaussian process $\{X_{uv}:u,v=0,\pm1,...\}$ and let

$$f(w,\eta) = \Sigma\Sigma r_{st}\exp\{-i(ws+\eta t)\} \tag{3.49}$$

denote a corresponding spectral density function, where ranges of summation are to be taken over all distinct lattice points. Such a process is said to be a finite order autonormal process on the infinite rectangular lattice (37) if

$$f(w,\eta) = \sigma^2\{1-\Sigma\Sigma\beta_{k\ell}\cos(wk+\eta\ell)\}^{-1} \tag{3.50}$$

where (i) only a limited number of the real coefficients $\beta_{k\ell}$ are nonzero, (ii) $\beta_{00}=0$, (iii) $\beta_{k,-\ell}=\beta_{k,\ell}$, (iv) σ^2 is a positive constant and (v) $\Sigma\Sigma\beta_{k\ell}\cos(wk+\eta\ell)<1$ for $0\leq w,\eta\leq 2\pi$. The existence of such schemes has been established by various methods, see Rosanov (32), Bartlett (38) and Moran (40).

The parameters of the scheme are related by the equations (3.50)

$$\sigma^2 = r_{00}-\Sigma\Sigma\beta_{k\ell}r_{k\ell} \tag{3.51}$$

Some simplifications can be achieved if we replace the infinite lattice autonormal scheme by its analogue on the pxq torus lattice. That is, we identify u with u+p and v with v+q, we need to only consider finite dimensional distributions associated with the set $\{X_{uv}: u=0,\dots p-1;\ v=0\ ,\dots\ q-1\}$. If we redefine the spectral density function as

$$g_{mn} = \sum_{s=0}^{p-1}\sum_{t=0}^{q-1} r_{st}\exp\{-i(w_m s+\eta_n t)\} \tag{3.52}$$

where $w_m = \frac{2\pi m}{p}$, $\eta_n = \frac{2\pi n}{q}$ and m,n are integers, then the auto normal scheme corresponding to that of infinite plane lattice has spectral function (3.50) at points where $w=w_m$ and $\eta=\eta_n$. Condition (i) is formally replaced by (ia) $\beta_{k\ell}=0$ unless $0\leq k\leq p-1$, $0\leq\ell\leq q-1$.

The joint probability function for the torus lattice variables is again

$$(2\pi\sigma^2)^{-pq/2}\ |B|^{1/2}\ \exp\{-\tfrac{1}{2}\sigma^{-2}(\underline{X}-\underline{u})^T B(\underline{X}-\underline{u})\} \tag{3.14}$$

$\sigma^2\underline{B}^{-1}$ is the dispersion matrix of the appropriately ordered X_{uv}'s. Following the argument of Moran (40) for the first-order schemes, it may be shown that

$$|\underline{B}| = \prod_{m=0}^{p-1}\prod_{n=0}^{q-1}\{1-\Sigma\Sigma\beta_{k\ell}\cos(w_m k+\eta_n\ell)\} \tag{3.53}$$

and that

$$r_{00}=(pq)^{-1}\sigma^{2}\sum_{m=0}^{p-1}\sum_{n=0}^{q-1}\{1-\Sigma\Sigma\beta_{k\ell}\cos(w_{m}k+\eta_{n}\ell)\}^{-1} \quad (3.54)$$

We will use maximum likelihood estimation (MLE) to obtain our estimates (36). By maximizing (3.14), or its logarithm

$$-\frac{1}{2}\,pq\,\log(2\pi\sigma^{2})+\frac{1}{2}\,\log|\underline{B}| - \frac{1}{2}\sigma^{-2}(\underline{X}-\underline{u})^{T}\underline{B}(\underline{X}-\underline{u}) \quad (3.55)$$

with respect to the unknown parameters, one obtains their MLE. Thus one may establish that $\hat{u} = \overline{\underline{X}}$, the sample mean, and that

$$\hat{\sigma}^{2} = (pq)^{-1}(\underline{X}-\underline{\hat{u}})^{T}\underline{\hat{B}}\,(\underline{X}-\underline{\hat{u}}) \quad (3.56)$$

Once $\underline{\hat{B}}$ has been found. If one writes

$$C_{st} = (pq)^{-1}\sum_{u=0}^{p-1}\sum_{v=0}^{q-1}(x_{uv}-\hat{u})\,(x_{u+s,v+t}-\hat{u}) \quad (3.57)$$

for the empirical auto-covariance corresponding to r_{st}, one finds that

$$\hat{\sigma}^{2} = C_{00} - \Sigma\Sigma\hat{\beta}_{k\ell}C_{k\ell} \quad (3.58)$$

and, substituting back into (3.55), $\hat{B}$ is obtained by minimizing

$$\text{Log}\,(C_{00}-\Sigma\Sigma\beta_{k\ell}C_{k\ell})-(pq)^{-1}\log|B| \quad (3.59)$$

with respect to the $\beta_{k\ell}$'s.

In practice, the quantity (3.55) may be minimized iteratively using (3.53) to determine $|B|$ at each stage. Newton-Raphson procedure is sometimes used.

For the specific first-order isotropic process on the torus lattice, it is possible to obtain analytical results on parameter estimation. In particular, consider the isotropic scheme for which

$$E(X_{uv}/\cdot) = u+\beta(x_{u-1,v}+x_{u+1,v}+x_{u,v-1}+x_{u,v+1}-4u) \quad (3.60)$$

and $|\beta|<\frac{1}{4}$. To obtain $\hat{\beta}$, we need to minimize

$$\log\{C_{00}-2\beta(C_{10}+C_{01})\}-(pq)^{-1}\sum_{m=0}^{p-1}\sum_{n=0}^{q-1}\log\{1-2\beta$$

$$(\cos w_m+\cos \eta_n)\} \quad (3.61)$$

with respect to β. In order to make analytical progress, it is convenient to replace summations by the appropriate approximating integrals. Writing (36)

$$V_{st}(\beta) = \frac{1}{4\pi^2}\int_0^{2\pi}\int_0^{2\pi}\frac{\cos sx \cos ty}{1-2\beta(\cos x + \cos y)}dxdy$$

We obtain $\hat{\beta}$ as the solution of

$$\frac{C_{10}+C_{01}}{C_{00}-2\beta(C_{10}+C_{01})} = V_{10}(\beta)+V_{01}(\beta)$$

or

$$V_{00}(\beta) = \frac{C_{00}}{C_{00}-2\beta(C_{10}+C_{01})} \quad (3.62)$$

Note that for the infinite lattice, we have $r_{st} = \sigma^2 V_{st}(\beta)$. The advantage of (3.62), as we have seen before, is that $V_{00}(\beta)$ is expressible in terms of the complete elliptic integral of the first kind.

A "coding" method of conditional parameter estimation for lattice Markov fields has been suggested by Besag (37). The appeal of the approach lies in its simplicity and in the availability of exact statistical tests concerning the resulting estimators.

In order to fit a first-order scheme, we begin by labeling the interior sites of the lattice, alternatively X and •, as shown in Figure 3.10. It is then immediately clear that, according to the first order Markov assumption, the variables associated with the

X sites, giving the observed values at all other sites, are mutually independent.

```
•  x  •  x  •  x  •  x
x  •  x  •  x  •  x  •
•  x  •  x  •  x  •  x
x  •  x  •  x  •  x  •
```

Figure 3.10 Coding Pattern for First Order Scheme.

This results in the simple conditional likelihood

$$\pi p_{i,j}(x_{ij}/x_{i-1,j}, x_{i+1,j}, x_{i,j-1}, x_{i,j+1})$$

for the x sites values, the product being taken over all x sites. Conditional MLE of the unknown parameters can be obtained in the usual way.

Thus, we consider an infinite regular lattice with sites now labeled by integers i. We suppose that each site has v neighbors. Let $\{X_i\}$ denote the associated stationary Gaussian scheme for which

$$E(X_i/\cdot)=u+\beta \sum_{j\varepsilon N(i)} (x_j-u) \quad var(X_i/\cdot)=\sigma^2$$

where $|\beta|<1/v$ and N(i) denotes the set of neighbors of site i. Now suppose we have available a partial realization of the process over a simply-connected region of N sites. The first step in the coding technique is to encode as many of these sites as possible, subject to the rules, first, that no two encoded sites are to be neighbors of each other and, second, that the neighbors of each encoded site must themselves belong to the region (36).

Suppose that under optimal coding, r sites, with labels $i_1,\ldots,i_r$, are encoded. Of course, r will, in general, depend upon N and upon the shape of the region. Write, for k=1,...r

$$U_k=X_{i_k}, \quad W_k = \sum_{j\varepsilon N(i_k)} X_j$$

Then, given $W = (W_1,\ldots,W_r)$, the variates $U_1,\ldots,U_r$

are mutually independent, with

$$E(U_k/W) = u+\beta(W_k-vu),$$

$$var(U_k/W) = \sigma^2$$

The coding estimator of β is defined as

$$\hat{\beta} = \sum_{k=1}^{r} U_k(W_k-\overline{W}) \;/\; \sum_{k=1}^{r} (W_k-\overline{W})^2 \qquad (3.63)$$

where $\overline{W} = (W_1+...+W_r)/r$. Clearly the estimator is unbiased, sine $E(\beta/W)=\beta=E(\beta)$, and this, in turn implies that

$$var\ (\beta) = E\ [var(\beta/W)] = E[\sigma^2/\Sigma(W_k-\overline{W})^2].$$

This coding technique is used later in finding our estimates in the experiment.

Now we have completed deriving the theoretical results necessary for our construction of contextual algorithm.

3.3.4 Normalization Procedure and Multivariate Site-Variable Consideration.

The developed stationary plane process in the previous section requires the variances of the site variables to be the same. In real situations, however, the site variables on the lattice are bound to have unequal variances. For example, consider a one-band LANDSAT data in which the different classes are characterized by different means and variances, thus causing data points on the two-dimensional plane to have different variances. In order to fit the stationary model to our data, a normalization procedure has to be employed to homogenize the data points such that they all have the same variance.

Consider the univariate site variable case first. Every pixel has a random variable X_{ij} associated with it. For convenience, we will also subtract the mean of the random variable X_{ij}. Then X_{ij} can be represented as

$$X_{ij} = M_{ij} + kE_{ij} \tag{3.64}$$

where E_{ij} has zero mean and unit variance.

Instead of considering the original random variable X_{ij}, we are now considering the normalized variate having zero mean and unit variance at every site on the lattice. Of course, the normalization requires the knowledge of the true classes at every pixel. This is the purpose of the preclassification process. The preclassification result will enable us to fit our model to real data.

What effect does the normalization procedure have on the contextual information which we employ? The correlations now are assumed to exist only among the error variates E_{ij}'s at different sites (instead of among the original X_{ij}'s). This means that the class-transition probability context is no longer used. Rather, the contextual information comes from the correlations between the normalized random variables on the lattice. These error variates associated with different sites can be thought of as the random errors generated when the LANDSAT data were being collected. Usually the random nature of the remotely sensed data is caused by instrumental scanning error, atmospheric turbulence, weather condition, etc. No matter what the causes are for the random deviation of the data, however, the effect that they have on different pixels should be the same because all data points within one image frame are taken at the same time. This implies that there does exist some correlation among the error variates at different sites, so our source of contextual information is justified.

In order to estimate the correlation parameter, the normalization procedure can be performed on the training data for which we know the true classes at every point. The contextual classification procedure then consists of the following:

A) Use an area of ground truth information to obtain the estimates of β by a coding technique (Equation (3.63) after the data points in the area are normalized). Obtain the estimate of σ^2 from Equation (3.58).

B) Use the simple decision rule (no context) to perform preclassification and store these results for later postclassification.

C) Use equations (3.44), (3.46), (3.48) to obtain appropriate variance and covariances (correlations after normalization) to formulate the joint pdf $P(x_k, x_{k_1}, x_{k_2}, x_{k_3}, x_{k_4}/\theta_k)$ for the five-dimensional Gaussian density function. The preclassified results are used here to evaluate $P(x_k, x_{k_1}, x_{k_2}, x_{k_3}, x_{k_4}/\theta_k)$.

D) In the contextual classification process, use the normalized pattern values $x_k, x_{k_1}, x_{k_2}, x_{k_3}, x_{k_4}$ to evaluate $P(x_k, x_{k_1}, x_{k_2}, x_{k_3}, x_{k_4}/\theta_k)$ for all possible values of θ_k. Calculate the posteriori probability $P(\theta_k/x_k, x_{k_1}, x_{k_2}, x_{k_3}, x_{k_4})$, and decide to which class site k belongs.

So far we have concentrated only on the single-variate site variable situation. Now we describe the solution to the multivariate site variable problem. A two-variate site variable case will be used to illustrate the idea, since more variates would merely be simple extensions of the two-variate solution.

When there are spatial relations between pairs of sites, the joint density function was used to incorporate these interactions. Now we have covariance between two spectral bands as well as the spatial interactions between pairs of sites. It is natural to form the joint density function of all the variates to incorporate all the correlations between them. There are the following kinds of correlations:

1. The spatial correlations between pairs of sites.
2. The correlation between pairs of variates at a particular site. These different variates at

each site correspond to different cnannels (spectral bands) for the satellite data.

3. The correlation between one variate of one site and another different variate of another site.

The first two kinds of correlation are already known. The third correlation must then be derived.

It would be ideal if we could obtain a multivariate site variable Gaussian Markov lattice process either by constructing it following the procedures used in the univariate case or by extending the univariate lattice process structure to a multivariate one. However, to our best knowledge, the theoretical existence of the multivariate Gaussian Markov process (in the sense of Eq. (3.2)) has not been verified. This can be shown by the following argument.

Consider the situation where, at each site of the lattice, there are two Gaussian random variates jointly distributed as shown in Figure 3.11. Suppose that this bivariate lattice process has the Markov property as defined in (3.2). Since the multivariate normal distribution is characterized by the property that every linear function of the variates has a univariate normal distribution (45), it is then required that the univariate lattice process formed by the linear combination of the two random variables at every site must also possess the Markov property (see Figure 3.12). The process in Figure 3.12 shall herein be called the resultant process. Equivalently, this requires that all the variances and covariances for the resultant process be evaluated from the lattices Green's function (3.44) for some correlation parameter value a'.

Suppose that the variates at every site have been normalized to have zero mean vector and identity covariance matrix. The variance for, say, site 1 in Figure 3.12 then becomes

$$c_1^2+c_2^2+2c_1c_2E[X_k,X_k'] \tag{3.65}$$

Therefore, the variance at any site ℓ of the resultant process is increased by a factor of $c_1^2+c_2^2+2c_1c_2E[X_\ell,X_\ell']$, since the original variances at every site was 1 (remember the ordering of sites in

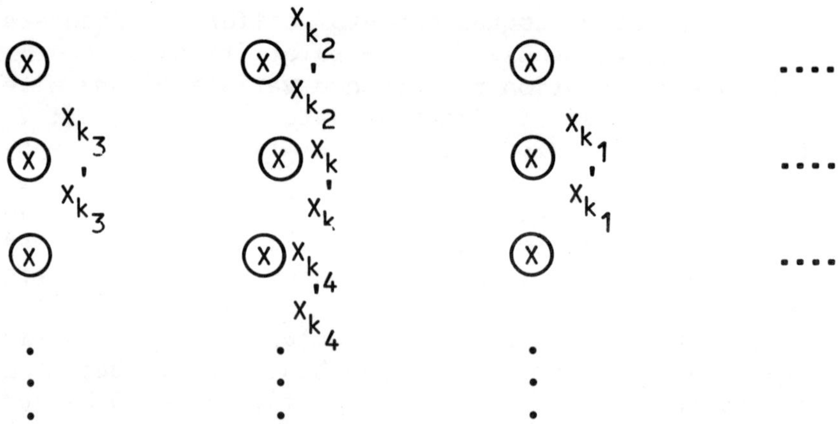

Figure 3.11 Two site-variable case for the NB system

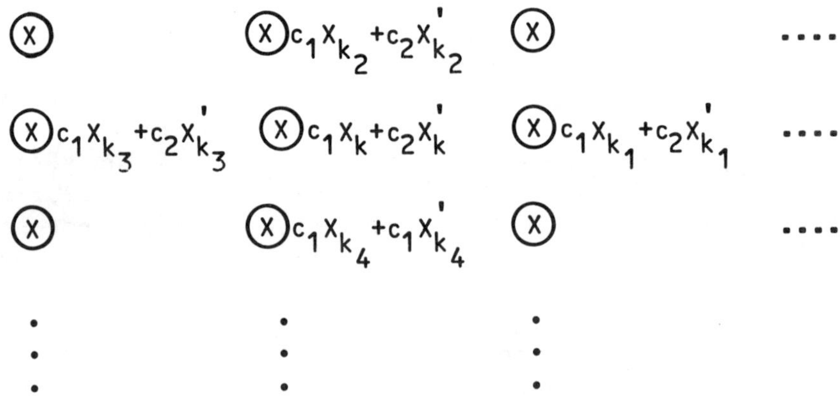

Figure 3.12 Univariate NB System Obtained from the Bivariate Process

Figure 2.1). Let's call this new variance V'_{00} where the prime here denotes the resultant process. In order for the resultant process to preserve Besag's Markov property, the covariance between site 1 and, say, site 2 must be (from equation (3.48))

$$V'_{10} = \frac{1}{4a'}(1-V'_{00}) = \frac{1}{4a'}[1-(c_1^2+c_2^2+2c_1c_2E[X_k, X'_k])] \tag{3.66}$$

On the other hand, the covariance V'_{10} can be computed from its definition. From Figure 3.12, we have

$$V'_{10} = E[(c_1X_k+c_2X'_k)(c_1X_{k_1}+c_2X'_{k_1}]$$

$$= c_1^2E[X_k,X_{k_1}]+c_2^2E[X'_k,X'_{k_1}]+2c_1c_2E[X_k,X'_{k_1}]$$

$$= E[X_k,X_{k_1}]\{c_1^2+c_2^2+2c_1c_2\frac{E[X_k,X'_{k_1}]}{E[X_k,X_{k_1}]}\}, \qquad (3.67)$$

since we have assumed that $E[X_k,X_{k_1}] = E[X'_k,X'_{k_1}]$ and $E[X_k,X'_{k_1}] = E[X'_k,X_{k_1}]$. By equating (3.66) to (3.67), we find that

$$E[X_k,X'_{k_1}] = \frac{1-(c_1^2+c_2^2+2c_1c_2E[X_k,X'_k]-4a'V_{10}(c_1^2-c_2^2)}{8a'c_1c_2} \qquad (3.68)$$

Clearly, the value of $E[X_k,X'_{k_1}]$ in (3.68) is a function of the real constants c_1 and c_2 if we enforce the resultant process to preserve the Markov structure. But the multivariate normal random vector definition requires that the resultant process be Markov for every linear function of the variates (i.e., $E[X_k,X'_{k_1}]$ should be independent from c_1 and c_2) if the multivariate site variable process were Markovian. Hence we have arrived at a contradiction. We can only hope to rely on assumptions and approximations to derive applicable results for the multivariate site variable situation.

In order to retain the Markov structure for the resultant process we must find some spatial correlation parameter a' (independent from c_1 and c_2) and use that in defining the corresponding torus process structure (3.32) and the infinite stationary Gaussian process (3.44). Consider the dispersion matrix of the torus process $\sigma^2 b_{st}$ (3.39) where σ^2 is the conditional variance in (3.13). When the size of the torus lattice process becomes large, $\sigma^2 b_{st}$ should be very close to the covariance V_{st} evaluated for the stationary lattice process. Using this approximation, (3.65) and (3.67) becomes $\sigma^2 b_{00}$ and $\sigma^2 b_{10}$ respectively. Suppose we require that all the variances and covariances for the resultant process to be constant multiples of the variance and covariances of the original univariate process, then the dispersion matrix for the torus process corresponding to the resultant process would be a constant, say c, times the original dispersion matrix. Equivalently, we are using the matrix

$$c \begin{bmatrix} 1 & a & o & o & . & . & . & o \\ a & 1 & a & o & . & . & . & o \\ o & a & 1 & a & . & . & . & o \\ . & . & . & . & . & . & . & . \\ a & o & . & . & . & o & a & 1 \end{bmatrix} = \begin{bmatrix} c & ca & o & o & . & . & . & o \\ ca & c & ca & o & . & . & . & o \\ o & ca & c & ca & . & . & . & o \\ . & . & . & . & . & . & . & . \\ ca & o & . & . & . & o & ca & c \end{bmatrix} \tag{3.69}$$

to approximate

$$\begin{bmatrix} 1 & ca & o & o & . & . & . & o \\ ca & 1 & ca & o & . & . & . & o \\ o & ca & 1 & ca & . & . & . & o \\ . & . & . & . & . & . & . & . \\ ca & o & . & . & . & . & ca & 1 \end{bmatrix} \tag{3.70}$$

From equations (3.65) and (3.67) we see that, in order for the two quantities to undergo the same scale change, the following condition

$$E[X_k, X_k^{'}] = \frac{E[X_k, X_{k_1}^{'}]}{E[X_k, X_{k_1}]} \qquad (3.71)$$

must be imposed. The other covariances like V_{11}, V_{20}, etc. will entail us to impose the same conditions, i.e., the type three correlation is equal to the product of the type 1 and type 2 correlations. It seems that the correlations between pairs of variates at every site would have to be the same for this result to be consistent. Fortunately, the orthonormalization procedure discussed below will make all correlations between variates at every site to be zero, and the results derived in this section and the stochastic model proposed will then be feasible.

For the normalization of a two or more dimensional pattern vector, say $\underline{X}$, an orthonormal transformation procedure must be employed. When the transformation is linear, it can be represented as $\underline{Y} = \underline{A}\ \underline{X}$ where Y is the normalized pattern vector and $\underline{A}$ is the transformation matrix. Let $\Phi = (\Phi_1\ \Phi_2\ \ldots\ \Phi_n)$ be the eigenvector matrix where n is the dimensionality of $\underline{X}$. Let

$$\Lambda = \begin{bmatrix} \lambda_1 & & 0 \\ & \ddots & \\ 0 & & \lambda_n \end{bmatrix}$$

be the diagonal matrix of eigenvalues. Let $\underline{I}$ be the identity matrix.

We can find a linear transformation to diagonalize

the covariance matrix in the new coordinate system. (Remember that all means have been subtracted.) This means that we can obtain uncorrelated random variables in general and independent random variables for normal distributions. The transformation matrix $\underline{A}$ is the transpose of the eigenvector matrix of the covariance matrix. The eigenvalues are the variances of the transformed distribution.

After applying the orthonormal transformation, we can add another transformation $\Lambda^{-1/2}$ that will make the covariance matrix equal to $\underline{I}$ (82). The transformation matrix is now $\Lambda^{-1/2}\,\underline{\Phi}^T$ and is called the whitening transformation or the whitening process. It is the whitening transformation that we need to perform in order to fit the stationary spatial stochastic model to multivariate data.

After the whitening transformation, the covariance matrix of all the variables in the NB system is particularly easy to construct because the correlations between spectral bands no longer exist. For the bi-variate site variable case, suppose that the covariance matrix of the variables in the NB system is $\underline{\Sigma}_1$ for the first band and $\underline{\Sigma}_2$ for the second band. The 10 x 10 joint density function of all variables in the NB system for both bands is the $\underline{\Sigma} = \begin{matrix} \underline{\Sigma}_1 & \underline{0} \\ \underline{0} & \underline{\Sigma}_2 \end{matrix}$

(refer to (3.67) after pattern vectors have undergone the whitening transformation. Since $\underline{\Sigma}_1$ and $\underline{\Sigma}_2$ are both positive definite, it can easily be shown that $\underline{\Sigma}$ is also positive definite (111, p. 275, Ex. 7.21).

One distinctive advantage of the whitening process is that the computation time for the quadratic form of the joint density function is much less than that which would be needed if the spectral bands were dependent. The inverse of $\underline{\Sigma}$ is simply

$$\underline{\Sigma}^{-1} = \begin{bmatrix} \underline{\Sigma}_1^{-1} & 0 \\ 0 & \underline{\Sigma}_2^{-1} \end{bmatrix}$$

In the next chapter, we shall use the developed contextual classification procedures to perform a series of experiments and present the results.

CHAPTER 4
EXPERIMENTAL RESULTS

Several experimental results are presented in this chapter in an effort to demonstrate the feasibility of the developed contextual classification algorithm. Experiments were all carried out on real MSS data. To further demonstrate the proposed approach in other areas of application, experiments were performed on infrared aircraft data for the detection of tanks and airplanes. It is believed that the contextual classification procedure will be superior to that of the no-context classification in any practical pattern classification problem.

The first section of this chapter discusses the single stage contextual classification algorithm and presents some experimental results. Then, a recursive classification algorithm is proposed in the second section. Results obtained from the recursive algorithm are shown to be better than those of the single-stage approach. Finally, the last section presents some contextual classification results using tank and airplane data.

4.1 Single Stage Contextual Classification

4.1.1 Preclassification and Postclassification

If we restrict our attention to the NB system, the contextual decision rule is, as before, to choose that class θ_k which maximize the posteriori probability $P(\theta_k/x_k, x_{k_1}, x_{k_2}, x_{k_3}, x_{k_4})$. Since the classes of the neighboring cells k_1, k_2, k_3, k_4 are unknown, a class-vector approach was adopted as a preliminary method to implement the decision rule. However, it was found that the class-vector approach was marred by a number of disadvantages. The two most noteworthy disadvantages are the additional computation time needed to search through the large number of class vectors and the unreliability of the frequency tabulation due to limited ground truth information. Another method that can be used to find the presumably true classes of the neighboring patterns is to preclassify these cells by using Bayes or maximum-likelihood decision rule. This stage of the contextual algorithm is called the preclassification stage. The contextual classification of cell k by maximizing $P(\theta_k/x_k, x_{k_1}, x_{k_2}, x_{k_3}, x_{k_4})$ after all the neighboring cells are preclassified is then called post classification. The combination of these two classification stages is called a single-stage contextual classification algorithm. In Section 4.1.2 we will present some experimental results using the single-stage contextual algorithm.

4.1.2. Experimental Results of Single Stage Contextual Classifcation.

Following the classification procedures for univariate and multivariate site cases discussed in Section 3.3.4, experiments were performed on the LANDSAT data of Tippecanoe County. The supervised classification result was obtained by professional analysts at LARS (4-band maximum likelihood decision rule) and is shown in Figure 4.1. There were 17 spectral classes found in unsupervised classification and they are found to correspond to seven information classes with the help of reference data such as aerial photography. Table 4.1 shows all seventeen classes and the information classes. The statistics of these seventeen classes are also given in Table 4.1. In order to examine the statistics before doing any classification, we obtained several plots which gave us some idea of the separabilities between classes. The first one is the cospectral plot for all found bands. This plot is formed by indicating each class mean and an interval of plus or minus one standard deviation from the mean (Figure 4.2). However this plot does not tell us very much about the orientation of the joint density function of two or more bands. Two other plots showing orientations of a combination of band 0.6-0.7 μM and band 0.7-0.8 μM were obtained and are given in Figures 4.3, 4.4. They are the means plot, and the ellipse plot (Figures 4.3 and 4.4). A preliminary examination of these plots tells us that all the agriculture classes can easily be separated using band 0.6-0.7 μM. The pastures are confused with agricultural areas in band 0.7-0.8 μM. but can be separated using band 0.6-0.7 μM. Forests can be separated from the other classes using both bands. Residential areas can, to a large extent, also be separated using both bands. Water is distinctively shown in band 0.7-0.8 μM. The classes which are difficult to classify are highway and commercial areas. We shall see that the contextual classification rule can more effectively detect these two classes than the simple (No-memory) classification rule.

Table 4.1 (A) 17 spectral classes and the 7 corresponding information classes.

Information Classes	Spectral Classes
Highway	Highway
Agriculture	AG2-1 AG2-2 AG2-3 AG2-4 AG2-6 AG2-7
Pasture	Past-Wht Past Past 2-11
Forest	For2-12 For2-13
Commercial Area	Coml-1
Residencial Area	Res1-3 Res1-7
Agriculture	Dark Ag
Water	Water

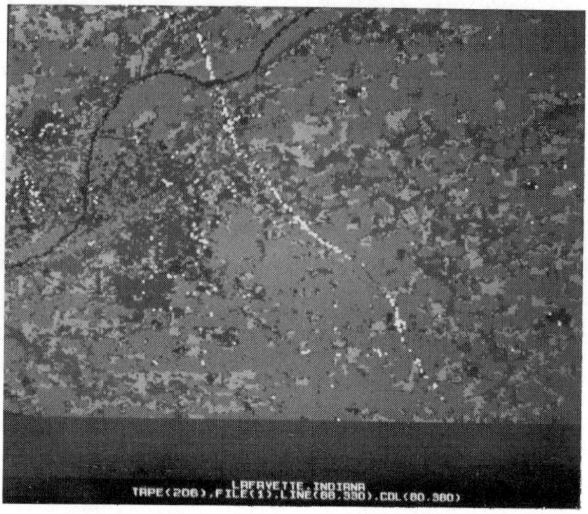

Figure 4.1. 4-band maximum likelihood classification rule result of Lafayette, Indiana.

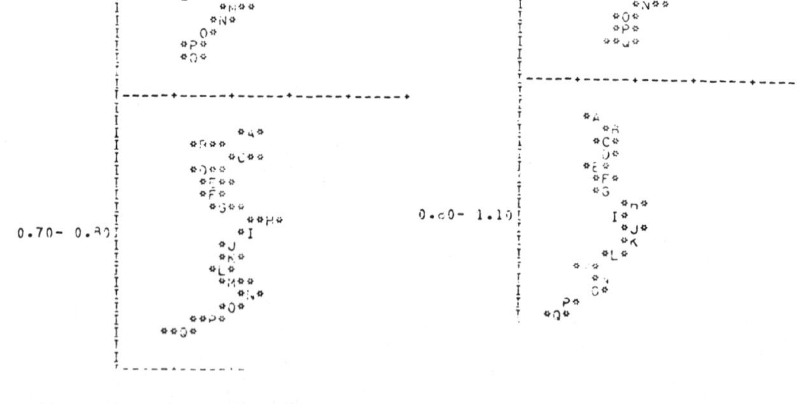

Figure 4.2. Cospectral plot of 17 classes (same as Figure 2.5).

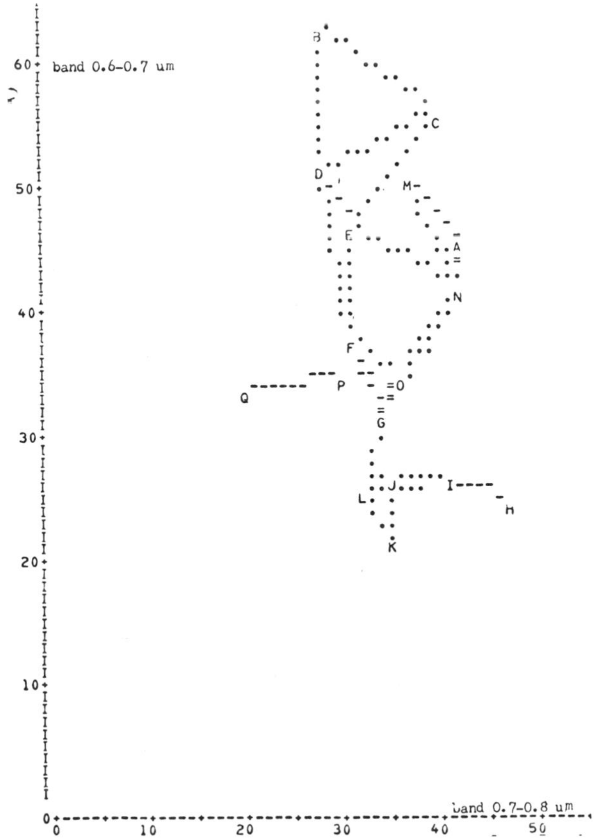

Figure 4.3. Means plot of band 0.7-0.8 um vs. 0.6-0.7 um for 17 classes of Lafayette data. Statistical distances between different means are also shown.

TRANSFORMED DIVERGENCE CODING

= 0000 TO 1000 - 1201 TO 1500

+ 1001 TO 1250 . 1501 TO 1750

. . . . .

Legend

SYMBOL	CLASS NAME
A	HIGHWAY
B	AG2-1
C	AG2-2
D	AG2-3
E	AG2-4
F	AG2-6
G	AG2-7
H	PAST-WHT
I	PAST
J	PAST2-11
K	FOR2-12
L	FOR2-13
M	COM1-1
N	RES1-3
O	RES1-7
P	DARK AG
Q	WATER

Figure 4.3. (Continued)

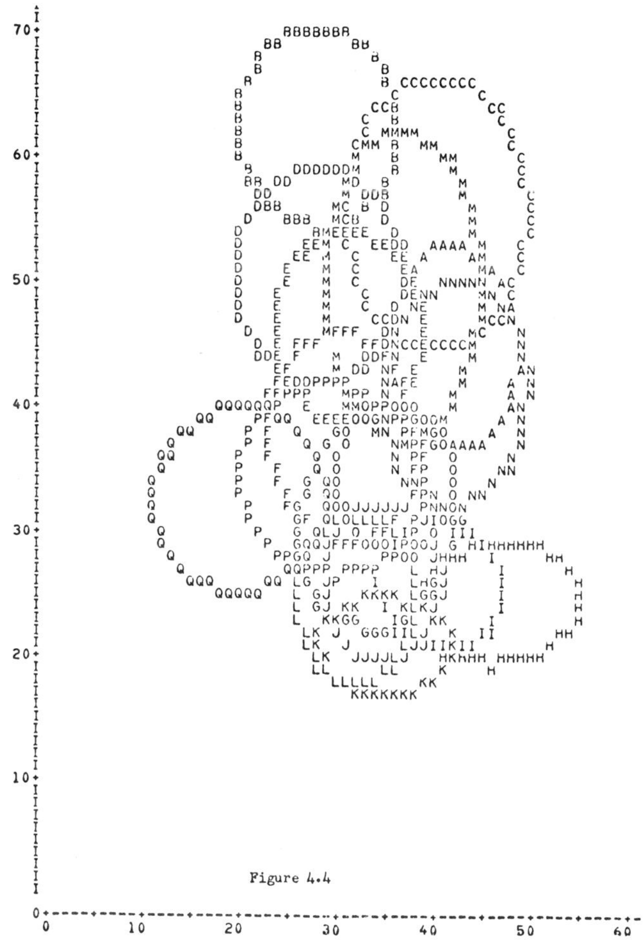

Figure 4.4. Ellipse plot of band 0.7-0.8 μM vs. 0.6-0.7 μM for 17 classes of Lafayette data.

(A) Preparation of the Reference Data

There are several reasons why the reference data are important. First the analyst has to gain some familiarity with the geographic region being analyzed. He can then associate the spectral classes found in unsupervised classification with the true information classes which are valuable to geologists. The reference data will also aid the analyst in developing good training statistics and

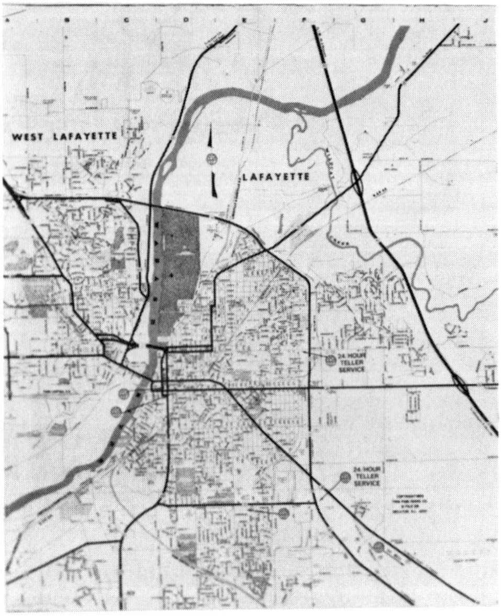

Figure 4.5. U.S.G.S. topographic map for Lafayette, Indiana.

in evaluating the classification results in the analysis process. There are several types of reference data which may be available to assist the analyst in his work. Typical examples are aerial photographs, maps (Figure 4.5), previous analysis results and on-site observations. One aid that is very useful is aerial photography. Four types of film are commonly used: black-and-white, black-and-white infrared, color, and color infrared. The type of film selected is determined by the nature of the application and the conditions under which it will be used. Infrared camera systems are often preferred for high altitude photography because of their haze penetration quality (61). Infrared films are also useful for enhancing differences between vegetation types. The color infrared aerial photography of Lafayette, Indiana area was obtained to provide ground truth information.

Although there was no photography taken at the time of data collection, we had available high altitude (18,000 M (60,000 ft)) color infrared photography taken in 1971, approximately one year prior to the collection of the 1972 LANDSAT data. While this reference data was not ideal, it was the best available. Maps are often used as reference data. The data can be preprocessed so that a computer line printer "map" has the same scale as the "topographic" map. A simple overlay technique could then be used to correlated the LANDSAT data with the map. By overlaying the computer printout of the area of interest on top of the quadrangel map, features of interest to the analyst can be located in the LANDSAT data and their row and column numbers noted (61). However it was found that a complete topo-map was not available and the Tippecanoe County LANDSAT data was not geometrically corrected for its skew and scale. Therefore only the aerial photography was used as reference data in our analysis.

(B) Analysis of Simple Classification Result

Before we deal with the contextual classification problem, all the statistics and simple classification results must be obtained in order to implement the contextual classification algorithm. Therefore the following procedures were used in the analysis of the results of simple classification. Training samples were selected from known urban areas and agricultural areas and the clustering algorithm (Figure 1.1) was employed to define spectral classes which would represent each general type of land use, i.e., urban, agriculture, pasture, forest, residential areas. The agricultural areas could not be identified as to specific crop since no concurrent photography was available, but this did not effect the analysis procedure since the goal was only to identify the areas in agricultural land use, essentially covered by green vegetation. The forest category and permanent pasture were also identified as a single group. After the cluster classes were identified in each individual area they were compared statistically and reduced in number as appropriate. For the 1972 Lafayette LANDSAT data, the final number of spectral classes used was 17.

Two areas of size 128 x 128 were classified with the no-context classification algorithm. Since it was necessary to test both univariate and multivariate contextual algorithms, one-band and two-band

a) 1-band simple rule result

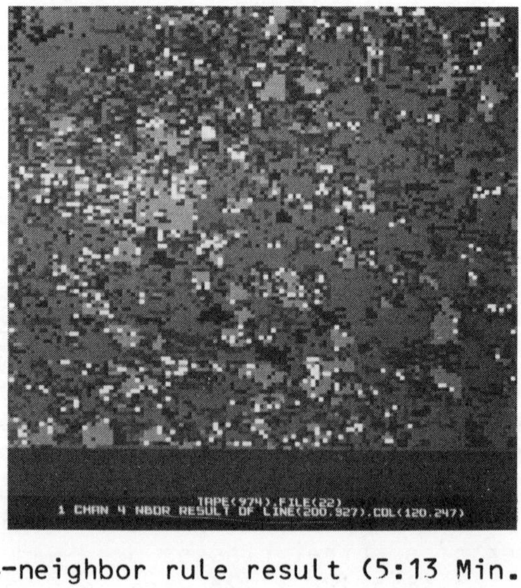

b) 1-band 4-neighbor rule result (5:13 Min. cpu time used)

Figure 4.6. 1-band classification result for block A.

a) 4-band simple result (10:00 Min. used)

b) 2-band simple result (2:37 Min. used)

Figure 4.7. Contextual results for Block A.

c) 2-band context result (9:18 Min. used)

Figure 4.7 (Continued)

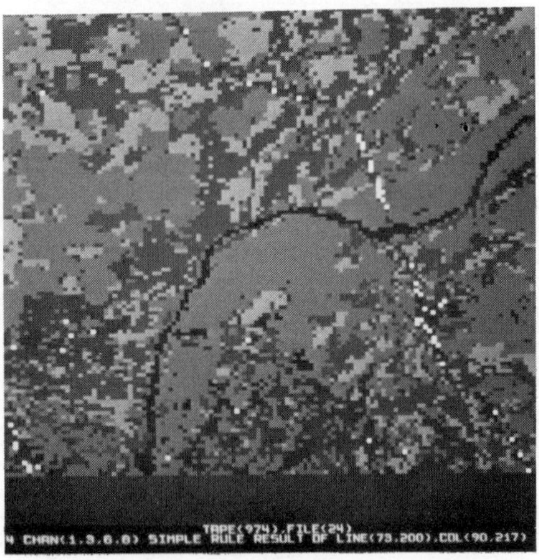

a) 4-band simple result (10:00 Min. used)

Figure 4.8. Contextual results for Block B.

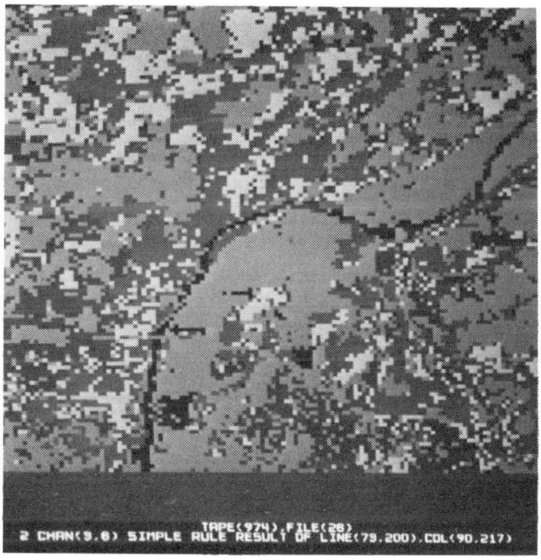

b) 2-band simple result (2:37 Min. used)

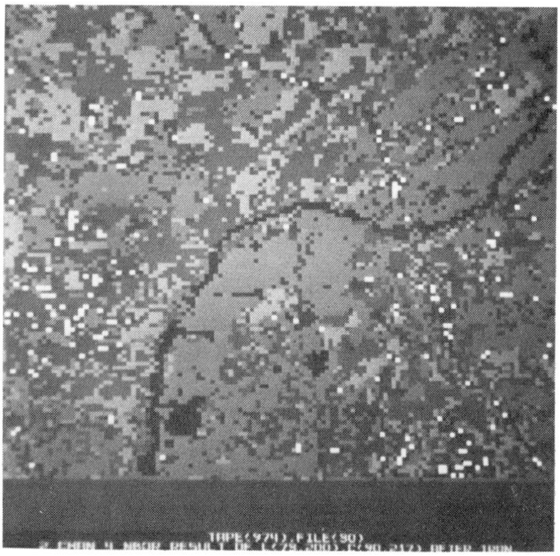

c) 2-band context result (9:18 Min. used)

Figure 4.8 (Continued)

classification results were both obtained for the two areas. The l-band result for Area A (line 200-327, col 120-247) is shown in Figure 4.6(a). The 2-band simple classification result of Area A is given in Figure 4.7(b) and that of Area B (lines 73-200, col. 90-217) is give in Figure 4.8(b).

Parameter Estimation by Coding Technique

In order to implement the contextual classification algorithm we have to estimate correlations (or covariances) between site variables as a function of distance. It is possible to estimate these correlations from a finite rectangular lattice. The ground truth information for the area used in the estimation must be available because we have to perform the normalization process, as discussed in Section 3.3.4. However the ground truth for any reasonably large field is not available for the following two reasons. First, we do not know exactly where the boundaries between different classes are, even with available aerial photography, because of the noisy nature of LANDSAT data. Secondly, it is an extremely complicated process to store these true classes for all sites in the field on tape or cards for users to access. The homogeneous areas for which we know the true classes are usually two small to be used for estimation. Therefore we adopt another approach to perform the estimation process.

Since the entire area of Lafayette has been classified by professional analysts using both supervised and nonsupervised techniques, we may take advantage of this fact by using the supervised classification result as the truth information for estimation. These classification results are stored automatically on result tapes and can be accessed easily by users. In this way, a reasonably large area can be used to simulate a rectangular lattice process which is a partial realization of an infinite plane process. The block area used to perform the estimation is Area C (lines 125-252, col. 187-314). The spatial interaction parameter β was estimated first by using the coding method, described in Section 3.3.3. The appropriate covariances between site-variables within the NB system were evaluated using (3.44). We used summations to replace the integrals in (3.44) with integration over the range from 0 to 2π divided into 100 intervals to approximate the lattice Green's function because the integrals are very hard to evaluate.

The results of estimation using the coding technique for two bands are given in Table 4.2.

Table 4.2 Coding Estimation Results of Spatial Corrrelation Parameter and Variance, Covariances Between Site Variables.

Band	$\hat{\beta}$	V_{00}	V_{01}	V_{11}	V_{02}
0.7-0.8 um	0.228	1.489	0.5372	0.3216	0.221113
0.6-0.7 um	0.229	1.486	0.5370	0.3212	0.2210

(D) Contextual Classification Results

After the estimation process, the contextual classification algorithm was tested using the LANDSAT data. First the preclassifications of areas A and B were performed and results stored on tape. The contextual algorithm using the joint probability density function to introduce spatial interactions was tested on areas A and B. The results of the 2-band contextual classifications for areas A and B are given in Figures 4.7(c) and 4.8(c) respectively. The 1-band context result for Area A is given in Figure 4.6(b). The computation time for classification of the entire area is also given in these figures.

For applications of classification of LANDSAT data, consideration of using only two spectral bands will be adequate for the following reason. The four spectral bands of LANDSAT data cover both the visible and infrared regions of the electromagnetic spectrum. But the first two bands cover approximately the same portion of the visible spectrum; the other two bands cover approximately the same infrared portion of the spectrum. This implies that we can select one band of the two from each region to achieve nearly the same results as using all four bands. This is also observed by examining the cospectral plots in Figure 4.2.

For comparison purposes, we also performed the 4-band classifications for areas A and B and the results are given in Figures 4.7(a) and 4.8(a) respectively.

(E) Evaluation and Analysis of Classification Results

One typical method of evaluating classification performance involves selecting coordinates for homogeneous areas which are called "test fields". These are selected by the analyst to represent known cover types in the analysis area. Some type of reference data is usually used to choose the test fields (60).

Another method for evaluating the performance is to use statistically random or stratified procedures to select areas over which to make actual ground observations to check the classification results. This procedure can be the most valuable method of checking accuracy of the classification, providing observations are made soon after the analysis. Photographs can also be useful but are subject to human interpretation error (60).

In the beginning an attempt was made to select homogeneous test fields with the help of aerial photography. However, it was later found that these fields are either extremely hard to find (since they have to be selected either from Area A or B) or they have to be extremely small (10 x 10 for example) in order to guarantee homogeneity. Therefore another approach was adopted to evaluate the classification results. We are mostly interested in comparing the 2-band contextual classification results with the 2-band and 4-band simple rule results.

Visual examination of the performances of the classification results with the help of aerial photography reveals that forests, pastures and residential areas were classified equally well for the 2-band context, 2-band, and 4-band simple rules. Water was classified with high accuracy in all three cases. As we mentioned earlier, highway and commercial areas are extremely hard to detect due to their rare occurrence (low prior probabilities) in the data. In order to compare the performances of classification results for highway, water, and commercial areas, a white graylevel is used for these three classes, and all the other classes are given the same dark graylevel to produce better contrast. Water is included for orientation purposes. The decoded images of three classification results for area block B is given in Figure 4.9(a),(b),(c) respectively. The results are compared with the aid of aerial photography. Highway and commercial areas are the classes which have been classified much better by the contextual algorithm, as can be seen from Figure 4.9.

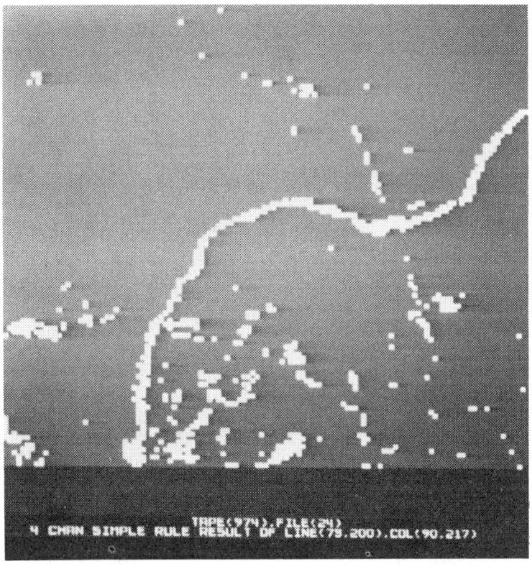

a) 4-band simple result

Figure 4.9. Decoded results for Block B.

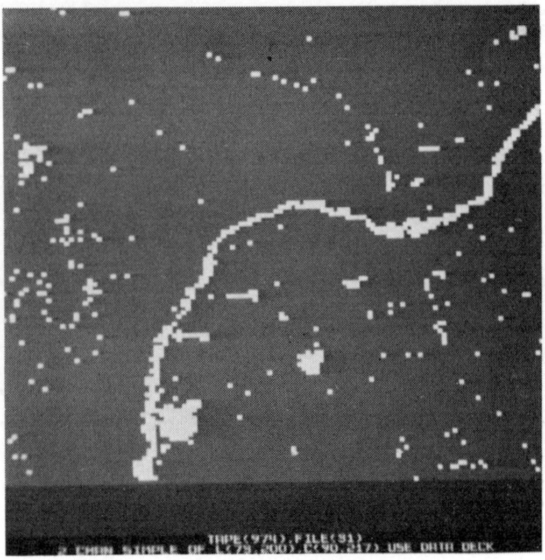

b) 2-band simple result

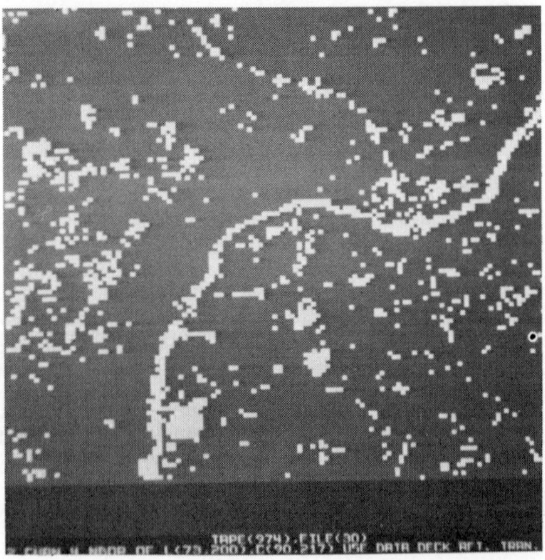

c) 2-band context result

Figure 4.9 (Continued)

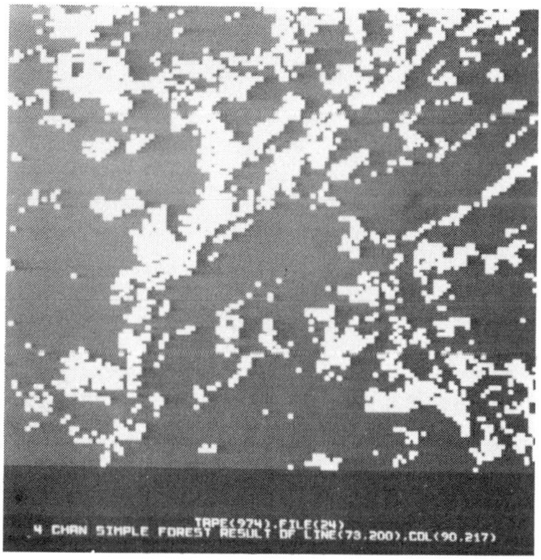

a) 4-band simple rule result

b) 2-band simple rule result

Figure 4.10. Decoded results for Class Forest.

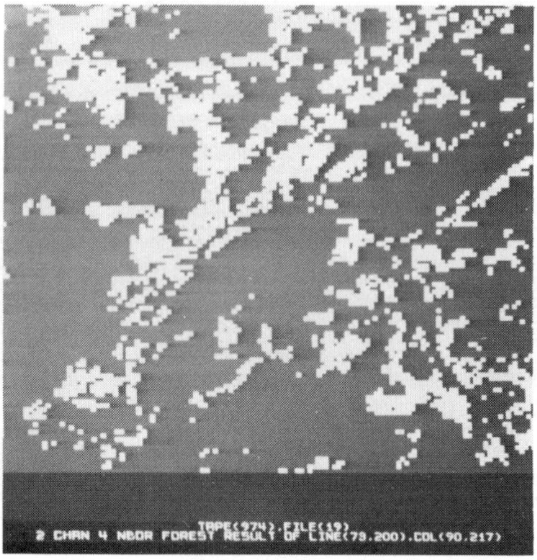

c) 2-band context rule result

Figure 4.10 (Continued)

a) 4-band simple rule result

Figure 4.11. Decoded results for Class Pasture.

b) 2-band simple rule result

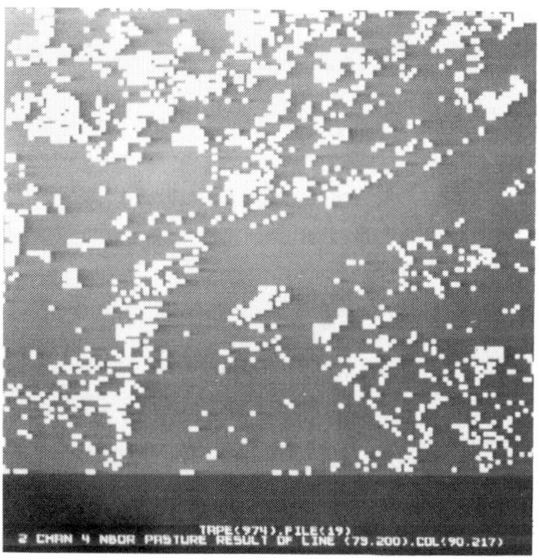

c) 2-band context rule result

Figure 4.11 (Continued)

The scattered white dots show the existence of commercial areas, and the aerial photography does support the presence of some commercial/industrial areas in this region. Even some city streets within the urban areas (to the left of the river in Figure 4.9(c) are detected by the contextual algorithm. The difference in the classification performances for highway and commercial areas is substantial. In order to compare the performances of other classes for the three classification rules, white graylevel was again used for classes forest and pasture, and the decoded results are shown in Figures 4.10 and 4.11 respectively. From these two figures, there does not seem to be much difference in performances for the three classification rules. Agricultural areas and residential areas are also found to be classified equally well by the three classification rules.

If a priori probabilities are good indication of the prior knowledge of the frequency of occurrences of individual classes, we can estimate the percentage improvement of the contextual result based on the above analysis. Since classes other than highway and commercial areas have been classified equally well, the difference between the contextual rule and the simple rule lies in the rarely occurring classes, highway and commercial areas. From the prior probability tabulation, the combined prior probability for highway and commercial areas is about 0.05. Therefore approximately 5% improvement in classification accuracy can be attributed to the use of contextual information. The computation time for the contextual algorithm is much less than that of the compound decision algorithm (refer to Figures 26 and 27) while the contextual algorithm can achieve nearly the same classification performance as the compound decision algorithm for the identical data set (Area A). In the following section we will investigate a multi-stage contextual classification scheme.

4.2 Recursive Contextual classification Results

4.2.1 Need of a Recursive Algorithm

To achieve the best possible performance with the contextual classifier, the true classes of the neighboring cells must be known with certainty. In almost all cases these classes are unknown to us. A preclassification technique was used in the previous section to obtain knowledgc about the neighboring classes and thereby to implement the contextual algorithm. Obviously the classes of the neighboring cells obtained by any kind of preclassification method will never be totally true and the subsequent contextual algorithm is therefore only suboptimal. On the other hand, the true classes of these neighbors are not known. This immediately suggests that perhaps by recursively classifying the patterns, one can gradually improve the knowledge of the neighboring classes and therefore also improve the contextual classification result. Eventually, the subsequent classification results will tend to remain unchanged and the recursive algorithm can then be stopped. In this way, we have compromised the above mentioned dilemma in arriving at a nearly optimal solution.

4.2.2 Recursive Classification Algorithm

Clearly an efficient mechanism is required to implement the recursive algorithm because of the increase in computation time associated with the recursive procedure. We could recursively classify all cells in the image by using the previous contextual classification result as the preclassification result. This is, however, not an efficient approach, because the true classes of these cells being contextually classified need not to be known. We can therefore save a considerable amount of computation time by first classifying only the neighboring patterns. Then, the cells whose neighboring classes are known can be classified contextually. We have yet to find an algorithm to implement this classification scheme.

Consider the coding patterns for the NB system shown in Figure 4.12. This coding pattern was used for the estimation of the spatial correlation parameter. However, it is also very useful for the recursive classification algorithm. Consider first the internal cells labelled with X in Figure 4.12. The neighboring cells of the X cells are the cells labeled with a dot. This implies that preclassification can be done first on the • cells in order to classify the X cells contextually. At this stage, the knowledge about the true classes of the • cells has been improved and the X cells can be contextually classified better than the previous time. This procedure can be carried out recursively. Only half of the data patterns are classified every time. When both the X cells and • cells are contextually classified once, we say that one stage of the recursive algorithm has been completed. The question now is when the recursive procedure should be stopped.

```
.  X  .  X  .  X  .  X  .  X
X  .  X  .  X  .  X  .  X  .
.  X  .  X  .  X  .  X  .  X
X  .  X  .  X  .  X  .  X  .
```

Figure 4.12 Coding Patterns for NB System.

The improvement of the knowledge about the neighboring classes will presumably improve the contextual classification result. If this is indeed the case (which will be shown by experimental results) the difference between successive contextual classification results sould be smaller and smaller as the recursive procedure goes on. Therefore, a good stopping criterion will be to examine the percentage difference between successive classification results. When the difference is smaller than a prespecified threshold the recursive procedure can be halted. The last stage of the recursive procedure becomes the final contextual classification result.

One should note that the final contextual classification result will not give 100% correct classification accuracy even if the ground truth is available. Rather it is the best result that can be obtained by using the spatial stochastic model. The lack of perfect fit of the model to the LANDSAT data due to the random nature of the data will inevitably cause erroneous classifications. However, it is the purpose of this investigation to demonstrate that

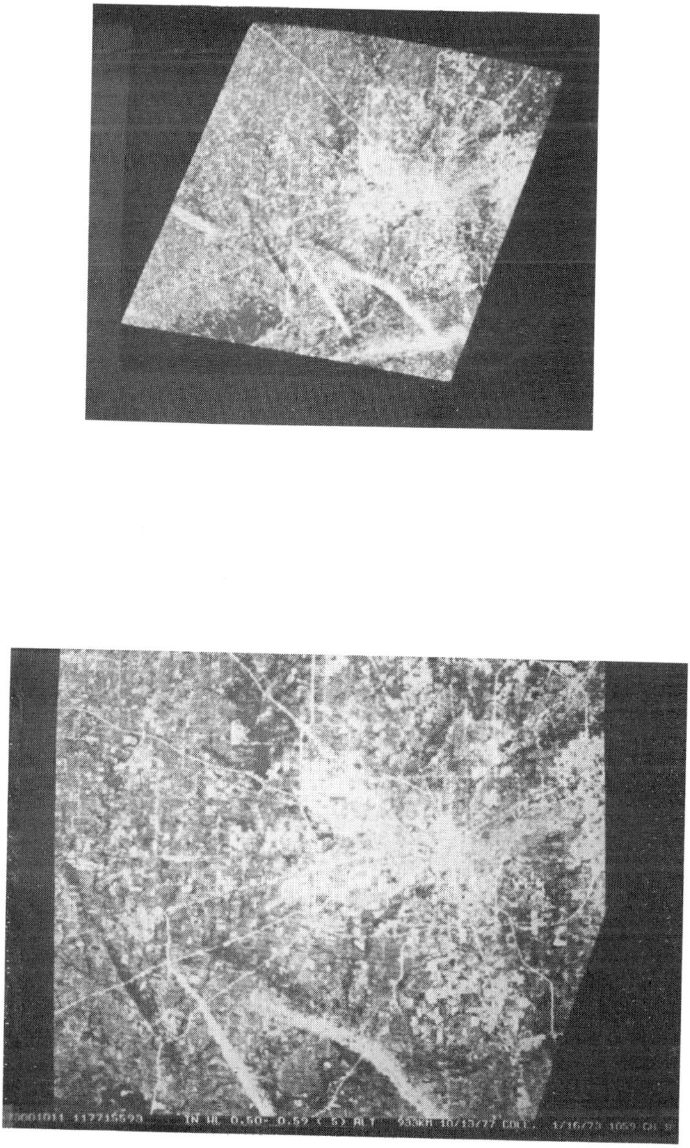

Figure 4.13. LANDSAT data of Marion County, Indiana.

the contextual classification model provides superior performance over the classical no-context classification. In the next section we will describe some experimental results of the recursive contextual classification algorithm.

4.2.3 Experimental Results of the Recursive Algorithm

Remotely sensed data collected from earth-orbiting satellites provide our experimental data. The Marion County (Indianapolis), Indiana LANDSAT data subframe was selected as the study area. Four bands of digitized, multispectral data from the ERTS (Earth Resources Technology Satellite) pass of 1972 were analyzed by the traditional simple classification rule and our recursive contextual algorithm in order to compare the performances.

The Marion County data were initially viewed on a digital image display (Figure 4.13) for orientation. Several small areas were defined for the clustering algorithm (43) (Figure 1.1). The clustering algorithm was used to find twenty-four clusters in the data. Those cluster classes were used to classify all the data points in the county, and the results were displayed by a line printer using different alphanumeric symbols for each class (60). The resulting cluster map of Marion County gave important clues to the spectral classes of urban land-cover phenomena, but a better definition of the land use classes was needed. Rectangular training samples of various land uses were located on the map, and their line and column coordinates recorded. Samples were located for the following land use: (1) commercial/industrial, (2) Golf course/cemeteries, (3) trees, (4) agricultural areas (5) older residential, (6) newer residential, and (7) water.

To illustrate the classification results with the map produced from the line printer is not practical because of its unwieldy size. Consequently the results were displayed on the digital image display (Figure 4.13). The simple rule (no context) two-band and four-band results together with the one-stage and two-stage recursive contextual rule results are given in Figure 4.15. Gray levels used

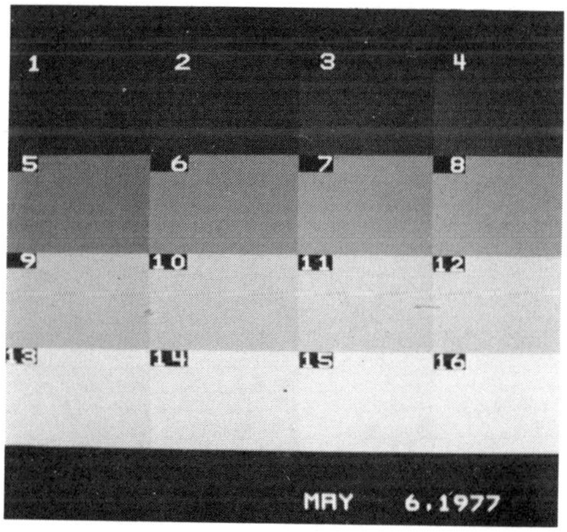

Figure 4.14. Gray level code for image display.

for the spectral classes are selected from the gray level code (Figure 4.14) as follows:

Classes	Gray Levels
Golf Course and Cemeteries	16
Trees	9
Agriculture Areas	7
Old Residential	5
New Residential	4
Commercial/industrial	3
Water	1

The statistics obtained for the twenty-four cluster classes are given in Figure 4.16. The ellipse plot for the bands 0.6-0.7 μM, 0.8-1.10 μM used to obtain the two-band simple result and all the contextual rule results are shown in Figure 4.17. From the cospectral and ellipse plots, one can easily tell that several classes are very hard to separate using training statistics. It can also be seen from these plots that the simple classification result obtained by using all four spectral bands will not

a) 2-band simple rule result

b) 4-band simple rule result

Figure 4.15. Recursive contextual classification results for area (Line 200-327, Col. 120-247) of Marion County data.

c) 1st stage context rule result

d) 2nd stage context rule result

Figure 4.15. (Continued)

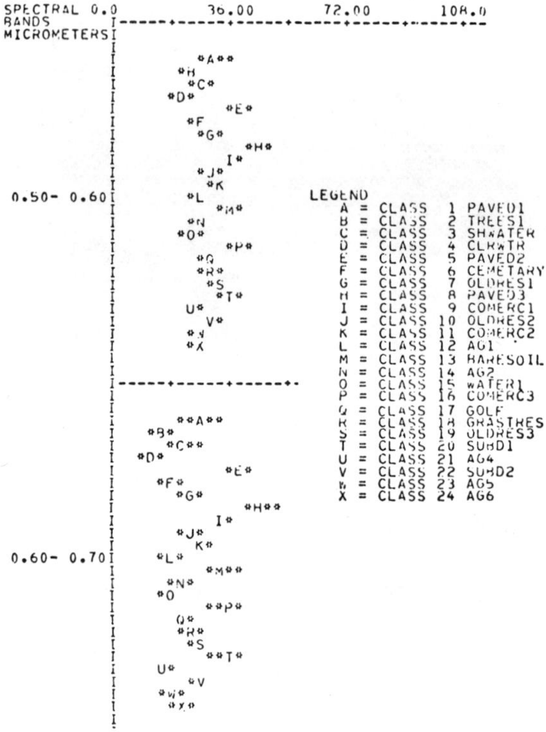

Figure 4.16. Cospectral plot for 24 classes of Marion County data.

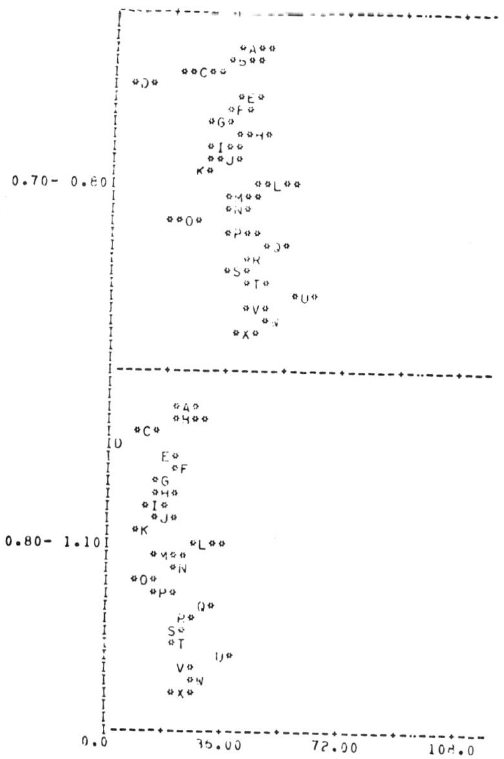

Figure 4.16. (Continued)

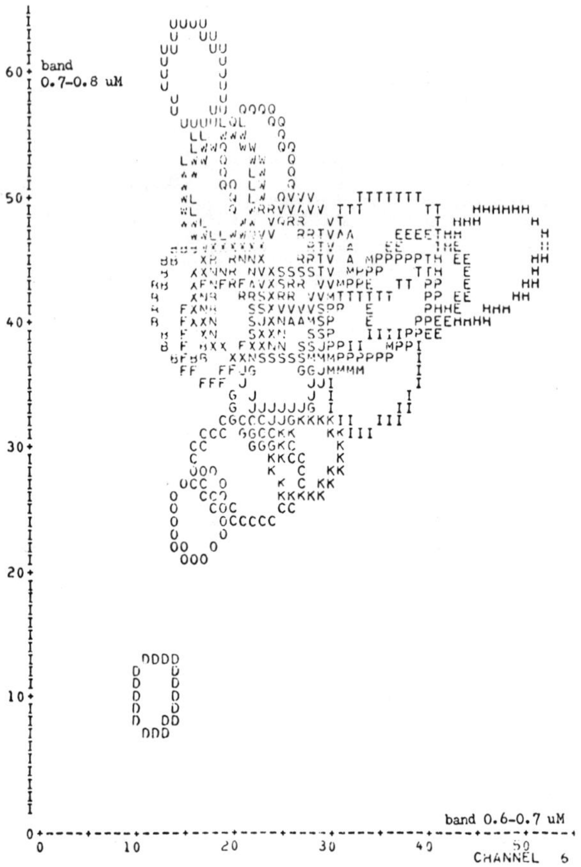

Figure 4.17. Ellipse plot for 24 classes of Marion County data.

be much different from that obtained by using only two bands. In fact, it was found that the performance of the four-band simple rule was worse than that of the two-band simple result because the classifier was confused by the extraneous information provided by the extra two bands. This will happen when the class means are very close to one another and the correlations between the spectral bands are low. Consequently the probability density functions will be distorted and the Gaussian assumption no longer holds. Examining the cospectral plots will show that this more or less is the case we encounter here. It is our intention then to demonstrate that the additional dimension of contextual information can improve the classification performance even after we have utilized the spectral information provided by all spectral bands.

The study area of size 128 x 128 was selected from the northwest portion of the county registered under LARS run number 73001011. The line and column ranges of this block are given in Figure 4.14. Several pairs of classes have been given visually similar graylevels, but consideration of their aerial distribution permits visual separation. For example, single-family (newer) residential and multi-family (older) residential areas are displayed as dark and medium gray. Commercial/industrial area is also displayed as dark gray, but visual separation is possible due to their different areal distribution.

The land use classes developed in this study are identifiable only because they consist of a unique type of ground cover or a unique combination of ground cover types. Four of the seven classes - trees, agriculture, water, golf course - are relatively homogeneous in character, i.e., they consist of a single land cover type. The other three classes - commercial areas, old residential and new residential - are mixtures of land cover types in various proportions. Rooftops and concrete such as highways are the primary constituents of the commercial/industrial class. The class multi-family (old) residential consists primarily of trees and rooftops. Finally, single-family (newer) residential has lawns (grass) and roads (concrete) as the principal components influencing its spectral nature.

Guidelines have been established for extracting land use information from remotely sensed data collected from space platforms (62). Table 4.3 compares that system with the one developed in this experiment. Level I of the Anderson et al. system (62) represents the land use categories which may be extracted from satellite data; information for levels II, III, and IV must be acquired from aerial photos and other sources. A close correspondence exists between the level I categories and the ERTS spectral classes.

Table 4.3 Comparison betweencclassification system proposed by Anderson, Hardy, and Roach and system used in the present study.

Anderson, Hardy and Roach System		Present Study
Level I	Level II	Spectral Class
01. Urban and Build-up Land		
	01. Residential	1. old 2. new
	02. Commercial & Services	Yes
	03. Industrial	
	04. Extractive	No
	05. Transportation, Communication, Utilities	No
	06. Institutional	No[1]
	07. Strip and Clustered Settlement	No[2]
	08. Mixed	No
	09. Open and other	Yes
02. Agriculture Land		Yes
03. Rangeland		NA[3]
04. Forest Land		Yes
05. Water		Yes
06. Nonforested Wetland		NA
07. Barren Land		NA
08. Permanent Snow & Icefields		NA

1 - Larger areas were classified as commercial/industrial

2 - Classified as new residential

3 - Not applicable to Marion County

It is difficult to obtain a quantitative expression of the accuracy of the Indianapolis classification results for several reasons. The most evident of these is the large number of data points which would have to be tested. The ground truth information for an area which covers more than one class is hard to obtain and extremely hard to store in the computer. Although the test fields for small rectangular homogeneous areas can be selected with the aid of reference data such as a U.S.G.S. topographic map, these small test-field performances by no means provide an overall view of the classification results because of the limited number of points that are tested. Also the performance along the boundaries between different classes is not known from the test-field performance. Even if an overlay of existing land use were to be employed for test purposes, a large number of data points would overlap different land uses (60). Another problem is the "training field accuracy" sample statistic which may be reported for each spectral class. Different land uses comprise various proportions of the total area from one metropolitan area to another. A sample size of 200 data points for the class commercial/industrial in two different counties may have very different weight, in consideration of overall classification accuracy, from one county to another (60).

With the above problems in mind, an attempt was still made to assess the classification accuracy by a sampling method. A U.S.G.S. topographic map was used as reference data to locate several rectangular areas, termed test fields (Table 4.4), for the spectral classes and the accuracies determined (Table 4.5). The first-stage recursive contextual classification result is about 5% better than the no-context classification result whereas the second-stage recursive contextual classification result is about 2% better than the 1st-stage result. The overall classification accuracy can reach as high as 95% by the recursive contextual rule. However, as we said before, these numeric accuracies do not give an overall view of the classification performances. It was then decided that we visually examine the

Table 4.4 Test fields for area (line 200-327, column 420-547) of Indianapolis data.

Pooled Classes	First Line	Last Line	Line Interval	First Column	Last Column	Column Interval
Test 1 (Golf) 73001011	251	253	1	516	521	1
Test 3 (Tree) 73001011	215	251	1	422	434	1
Test 2 (AG) 73001011	288	292	1	483	492	1
Test 4 (Old Res) 73001011	317	324	1	531	533	1
Test 5 (Comm) 73001011	253	259	1	525	528	1
Test 6 (New Res) 73001011	294	314	1	493	507	1
Test 7 (Water) 73001011	250	310	1	438	446	1

Table 4.5 Classification accuracies for area (line 200-327, column 420-547) of Indianapolis data.

Group	No of Samps	Pct Corct	Number of Samples Classified Into						
			Golf	AG	Tree	Old Res	Comm	New Res	Water
1 Golf	18	88.9	16	0	2	0	0	0	0
2 Ag	50	82.0	3	41	0	0	0	6	0
3 Tree	481	86.3	49	16	415	0	0	1	0
4 Old Res	24	87.5	0	0	0	21	3	0	0
5 Comm	28	60.7	0	1	1	1	17	8	0
6 New Res	315	81.9	0	3	10	28	16	258	0
7 Water	549	98.2	0	0	1	3	6	0	539
Total	1465		68	61	429	53	42	273	539

Overall Performance (1307/1465) = 89.2

Average Performance by Class (585.5/7) = 83.6

a) 2-band simple result

Table 4.5 (Continued)

Group	No of Samps	Pct Corct	Golf	AG	Number of Samples Classified Into Tree	Old Res	Comm	New Res	Water
1 Golf	18	44.4	8	7	2	0	0	1	0
2 Ag	50	88.0	0	44	0	0	0	6	0
3 Tree	481	82.1	55	30	395	0	0	1	0
4 Old Res	24	79.2	0	1	0	19	3	0	1
5 Comm	28	60.7	0	2	0	1	17	8	0
6 New Res	315	79.7	0	10	8	31	15	251	0
7 Water	549	98.2	0	1	0	2	7	0	539
Total	1465		63	95	405	53	42	267	540

Overall Performance (1273/1465) = 86.9

Average Performance by Class (532.3/7) = 76.0

b) 4-band simple result

Group	No of Samps	Pct Corct	Golf	AG	Number of Samples Classified Into Tree	Old Res	Comm	New Res	Water
1 Golf	18	83.3	15	1	2	0	0	0	0
2 Ag	50	84.0	1	42	2	0	0	5	0
3 Tree	481	96.3	6	11	463	0	0	1	0
4 Old Res	24	83.3	0	1	0	20	1	2	0
5 Comm	28	64.3	0	1	1	1	18	7	0
6 New Res	315	89.2	1	9	10	12	2	281	0
7 Water	549	99.3	0	0	1	2	1	0	545
Total	1465		23	65	479	35	22	296	545

Overall Performance (1384/1465) = 94.5

Average Performance by Class (599.7/7) = 85.7

c) 1st stage context result

Table 4.5 (Continued)

Group	No of Samps	Pct Corct	Number of Samples Classified Into Golf	AG	Tree	Old Res	Comm	New Res	Water
1 Golf	18	83.3	15	1	2	0	0	0	0
2 AG	50	82.0	1	41	3	0	0	5	0
3 Tree	481	97.5	3	8	469	0	0	1	0
4 Old Res	24	83.3	0	1	0	20	0	3	0
5 Comm	28	64.3	0	1	1	1	18	7	0
6 New Res	315	91.4	1	8	10	7	1	288	0
7 Water	549	99.3	0	0	1	2	1	0	545
Total	1465		20	60	486	30	20	304	545

Overall Performance (1396/1465) = 95.3

Average Performance by Class (601.2/7) = 85.9

d) 2nd stage context result

classification results through a digital image display.

Figure 4.15 shows the classification results displayed by using different gray levels representing different classes. Therefore, by visually examining these results, one can tell easily how good the performances are within each class, and also along the boundaries between classes. At first sight, we see that the simple rule results are quite noisy. First- and second-stage contextual rule results served to "clean up" the picture significantly. It is seen that many white dots (representing golf courses and cemeteries) were eliminated in the contextual rule results leaving the suburban areas much more homogeneous. Careful examination of the U.S.G.S. topographic map has confirmed that the suburban areas do not have as many golf courses and cemeteries as shown by the simple-rule results.

The performances of the residential area (lower right dark section) of the four results are about the same. Also the performances for highway, commercial areas and water (Eagle Creek at the left of the picture) are almost the same.

The second stage contextual rule result shows even more homogeneity within classes than the first stage result. This is expected because the spatial model is fitted better after the neighboring true classes knowledge has been refined by the first stage result.

In summary, the contextual rule performed better in homogeneous areas and along boundaries since spatial correlation is used. Supposing that there is a bad (deviated) sample within some homogeneous area, this pattern will inevitably be misclassified by the simple decision rule. But the normalized pattern has a good chance of being correctly classified if the correlations with its neighbors are considered. This is the reason why contextual rule results are more homogeneous.

To further demonstrate the difference between the simple rule results and the contextual rule results, we have established thresholds for the graylevels for different classes such that golf courses and cemeteries are represented by white graylevel and all other classes by dark graylevel as background. The results of these thresholded graylevels are shown in Figure 4.18. We can easily see from this figure that the dramatic elimination of the incorrectly classified golf course and cemeteries in the recursive context results.

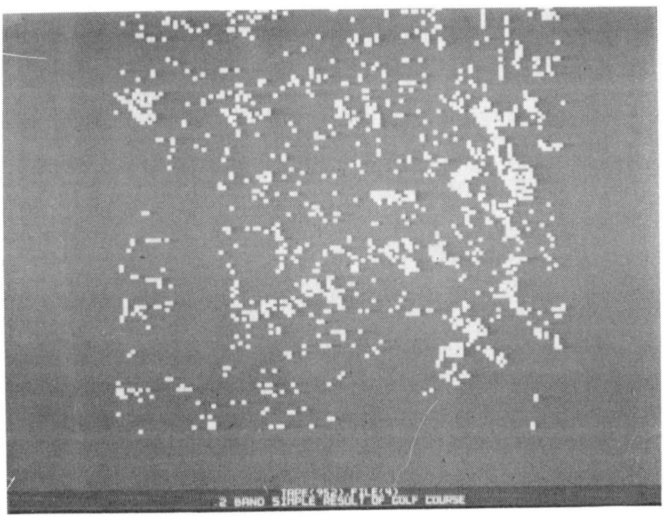

a) 2-band simple result

Figure 4.18. Thresholded results for golf courses.

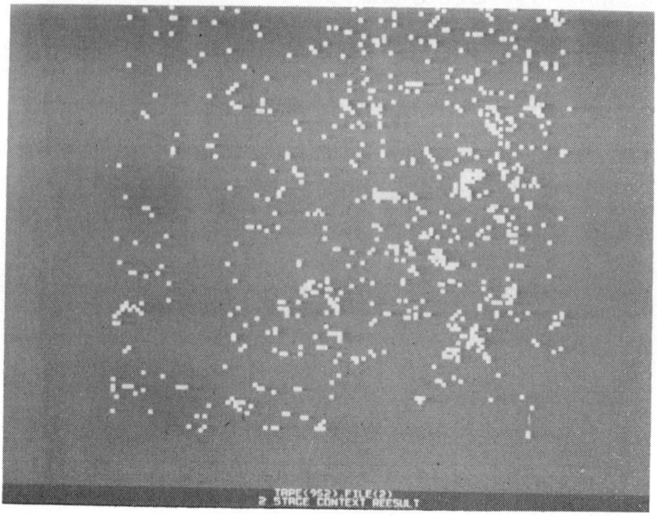

b) 1-stage context result

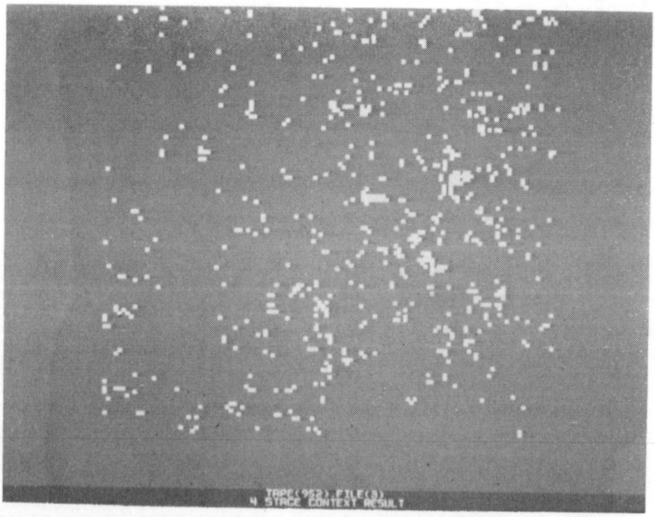

c) 2-stage context result

Figure 4.18 (Continued)

The first-stage context result is different from the result of the two-band simple rule by about 60%. The second-stage context result is different from the first stage by about 20% and the third-stage result is different from the second-stage result by less than 10%. Therefore the recursive contextual classification algorithm converges rather quickly. Since the improvement in classification accuracy is not significant after the second stage, it is believed that, for most LANDSAT data, two-stage classification will be enough for practical purposes. This means that the recursive algorithm is feasible because there is not much increase in computation time. The exact computation times are given in Figure 4.15. However, one should use his own judgement to decide whether it is worthwhile to continue the recursive process to obtain better classification results.

In order to gain information as to how well the recursive contextual algorithm performs within the metropolitan area, we selected another area block of size 128 x 128. The area was also registered under LARS run No. 73001011 with lines from 300-427 and columns from 550-677. White river passes through this area from north to south. There are mostly residential areas to the left and right of the river with some commercial area in the lower portion of this area. Agricultural areas and trees are the rarely occurring classes in this area.

The two-band and four-band simple rule results are shown in Figure 4.19. Results of three stages of the recursive contextual classification algorithm are given in Figure 4.20. We found the reason why this area took three stages of recursive algorithm to converge is because of the noisy nature of the data. Most of this area was covered by residential areas. However, the old and new residential areas were easily confused with other classes. Old residential areas are characterized by trees and streets since trees were grown up when the data was taken early fall in 1973. New residential areas were characterized, however, by rooftops (commercial areas) and grass. If we examine the cospectral plots for those classes, we find that the statistics of trees and old-residential areas are very similar,

a) 2-band simple rule result (7:00 Min. used)

b) 4-band simple rule result (28:00 Min. used)

Figure 4.19. Simple rule classification results for area (Line 300-427, Column 550-677) of Marion County (IBM 370 computer was used for this experiment).

a) 1st stage context (18:00 Min. used)

b) 2-stage context result (18:00 Min. used)

Figure 4.20. Recursive contextual classification results for area (Line 300-427, Column 550-677) of Marion County.

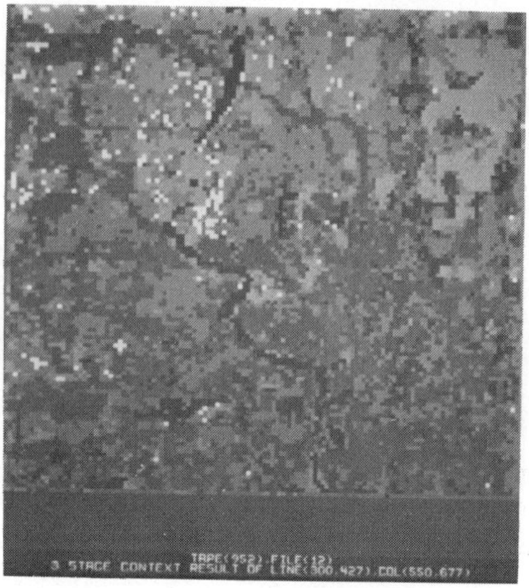

c) 3rd stage context result (18:00 Min. used)

Figure 4.20 (Continued)

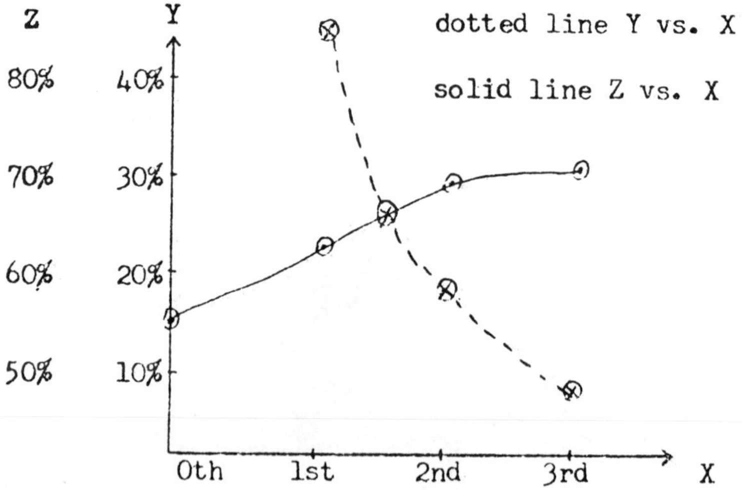

Figure 4.21. Percentage difference of successive stage (Y) vs. No. of stages and percentages of accuracy (Z) vs. X.

and the same is true for commercial areas and new residential areas. Therefore the recursive algorithm needs to take one more stage than in the previous experiment. Figure 4.21 gives the convergence rate for this experiment.

A landuser's main interest in classifying urban areas is to ascertain how much land is used for housing and how much is used for agriculture. Therefore, in the evaluation of classification accuracy, we pooled the two residential areas together, and also the agricultural and forested areas together (called Agtree). Small test fields (Table 4.6) were selected with reference to U.S.G.S. topographic map to assess the classification accuracy. Table 4.7 give the test field performances of all classification results. The percentages of accuracy versus number of stage's denoted by the solid line, is plotted also in Figure 4.21. The 0th stage represents the 2-band simple rule result.

Table 4.6 Test fields for area (line 300-427, column 550-677) of Indianapolis data.

Pooled Classes	First Line	Last Line	Line Interval	First Column	Last Column	Column Interval
Test 1 (Res)	379	390	1	646	651	1
73001011	375	590	1	615	636	1
Test 2 (Comm)	403	409	1	613	620	1
73001011	384	394	1	639	645	1
Test 3 (AG Tree)	355	361	1	566	577	1
73001011	370	382	1	584	594	1
Test 4 (Water)	315	326	1	601	603	1
73001011	369	370	1	599	605	1
	404	408	1	586	593	1
Test 5 (Golf)	354	360	1	593	596	1
73001011	350	353	1	596	596	1

Table 4.7 Classification accuracies for area (line 300-427, column 550-677) of Indianapolis data.

Group	No of Samps	Pct. Corct	Number of Samples Classified Into Res	Comm	Agtree	Water	Golf
1 Res	424	47.4	201	193	24	6	0
2 Comm	133	72.9	7	97	5	24	0
3 Agtree	227	45.4	120	2	103	0	2
4 Water	90	77.8	2	14	4	70	0
5 Golf	65	67.7	1	0	20	0	44
Total	939		331	306	156	100	46

Overall Performance (515/ 939) = 54.8

Average Performance by Class (311.2/5) = 62.2

a) 2-band simple result

Group	No of Samps	Pct. Corct	Number of Samples Classified Into Res	Comm	Agtree	Water	Golf
1 Res	424	53.8	228	80	94	22	0
2 Comm	133	56.4	7	75	20	31	0
3 Agtree	227	79.3	46	0	180	0	1
4 Water	90	80.0	5	2	11	72	0
5 Golf	65	50.8	1	0	31	0	33
Total	939		287	157	336	125	34

Overall Performance (588.939) = 62.0

Average Performance by Class (320.2/5) = 64.0

b) 1st stage context result

Table 4.7 (Continued)

Group	No of Samps	Pct. Corct	Number of Samples Classified Into Res	Comm	Agtree	Water	Golf
1 Res	424	64.4	273	47	89	15	0
2 Comm	133	50.4	21	67	22	23	0
3 Agtree	227	86.8	27	2	197	0	1
4. Water	90	86.7	1	0	11	78	0
5 Golf	65	46.2	1	0	34	0	30
Total	939		323	116	353	116	31

Overall Performance (645/939) = 68.7

Average Performance by Class (334.4/5) = 66.9

c) 2nd stage context result

Group	No of Samps	Pct. Corct	Number of Samples Classified Into Res	Comm	Agtree	Water	Golf
1 Res	424	69.1	293	38	82	11	0
2. Comm	133	43.6	20	58	34	21	0
3. Agtree	227	88.5	24	1	201	0	1
4. Water	90	86.7	1	0	11	78	0
5. Golf	65	46.2	1	0	34	0	30
Total	939		339	97	362	110	31

Overall Performance (660/939) = 70.3

Average Performance by Class (334.1/5) = 66.8

d) 3rd stage context result

Examination of Table 4.7 shows that both the accuracies in residential areas and agricultural areas increase as we recursively use contextual information. The accuracy decreases in the commercial area probably because there are not too many commercial areas in the data and the selected test fields may not be very accurate. The performances of water and golf course remain about the same.

To compare the classification performance of the residential area, we again thresholded the gray levels such that all residential areas are represented by white gray level and all other classes by dark gray level as background. Figure 4.22 shows the residential area results. As we can see from Figure 4.22 1st stage context result looks more or less the same as the simple rule result. At the end of three recursive stages of using contextual information, the contextual algorithm result (Figure 4.22d) looks much more homogeneous showing that residential areas have been clustered together.

Figure 4.23 shows the thresholded commercial area results. Many of the commercial areas were classified as residential area because of sites spectral similarity to old-residential area. Comparing the results with the topographic map does not show so much commercial area as shown in Figure 4.23(a). However, it does not show so few commercial area as in Figure 4.23(d) either. This is why the performance of commercial area for all results are not as good as other classes. However the overall performance of the recursive algorithm is much better than the no-context results. Computation time is three times longer than the simple classification algorithm for 1-stage contextual algorithm, six times longer for the 2-stage algorithm and so on. The exact computation times are given in Figure 4.20. Even though three recursive stages are needed for some data, the computation time is still less than ten times that of the simple decision rule.

4.3 Other Experimental Results

The flexibility of the contextual classification algorithm enhances its applicability to a wide range of problems to which the conventional no-context

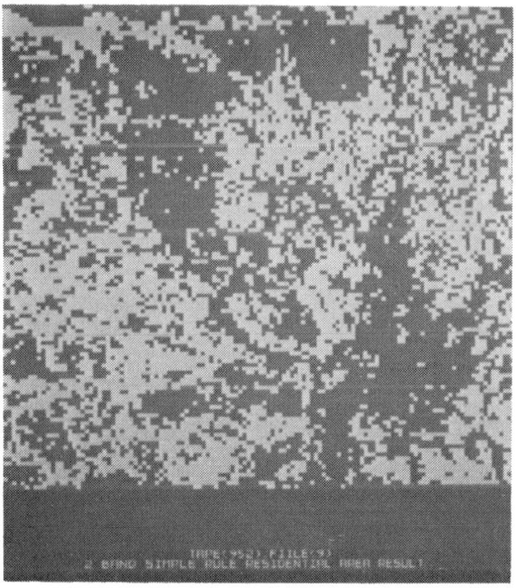

a) 2-band simple residential result

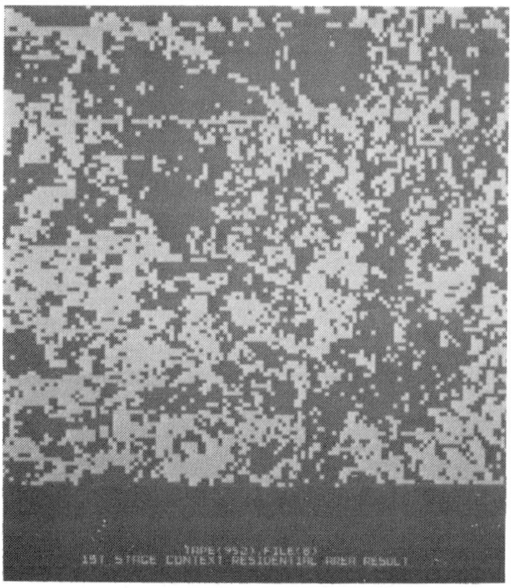

b) 1st stage context residential result

Figure 4.22. Thresholded results for Residential area.

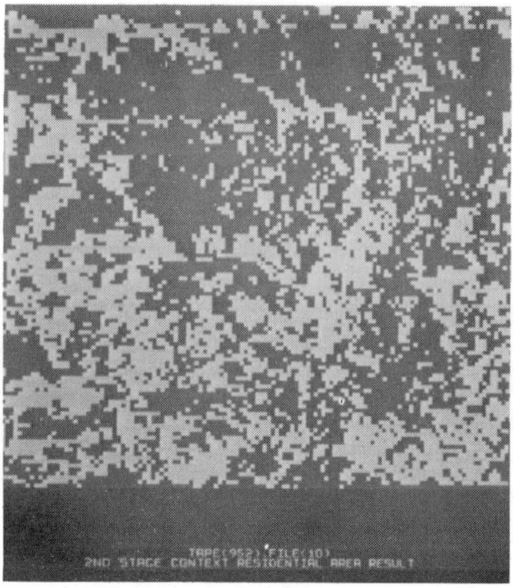

c) 2nd stage context residential result

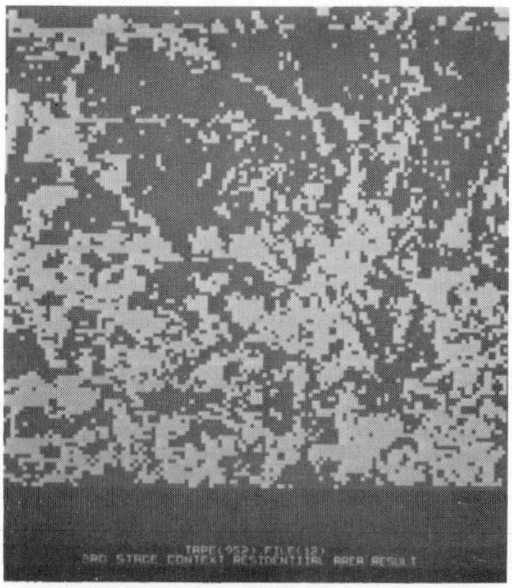

d) 3rd stage context residential result

Figure 4.22. (Continued)

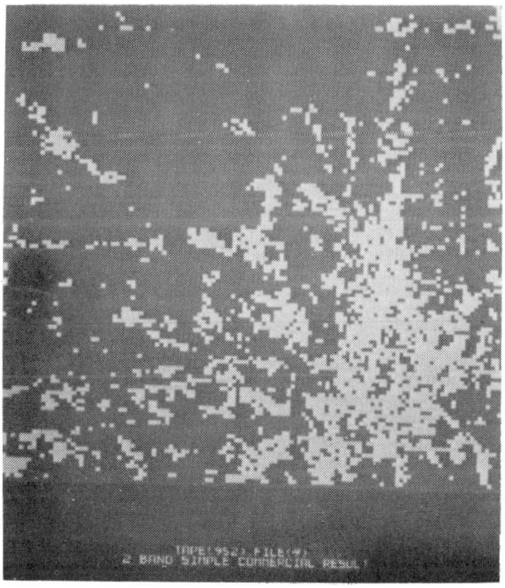

a) 2-band simple commercial area result

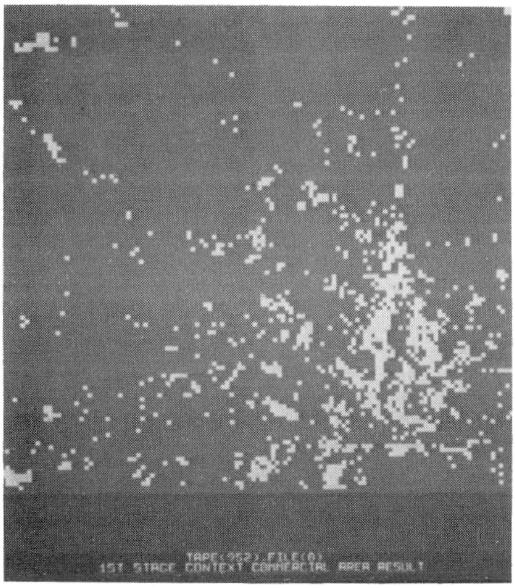

b) 1st stage context commercial area result

Figure 4.23. Thresholded Commercial area result.

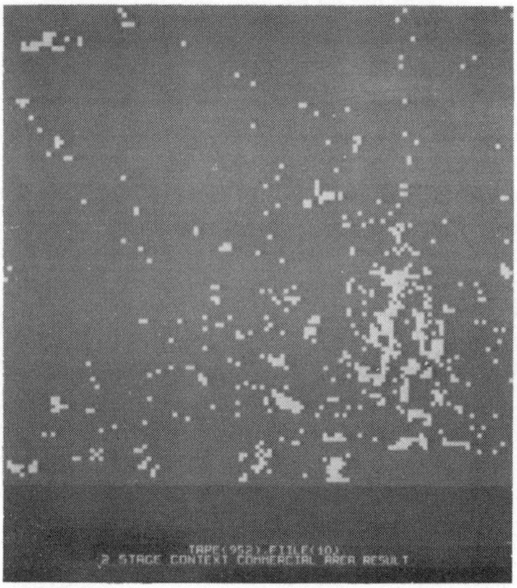

c) 2nd stage context commercial area result

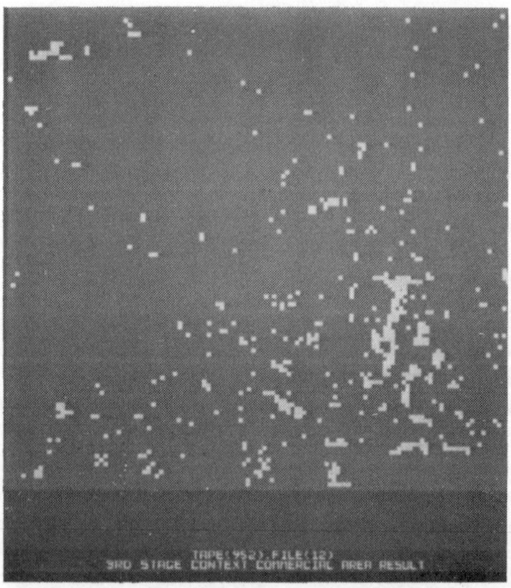

d) 3rd stage context commercial area result

Figure 4.23. (Continued)

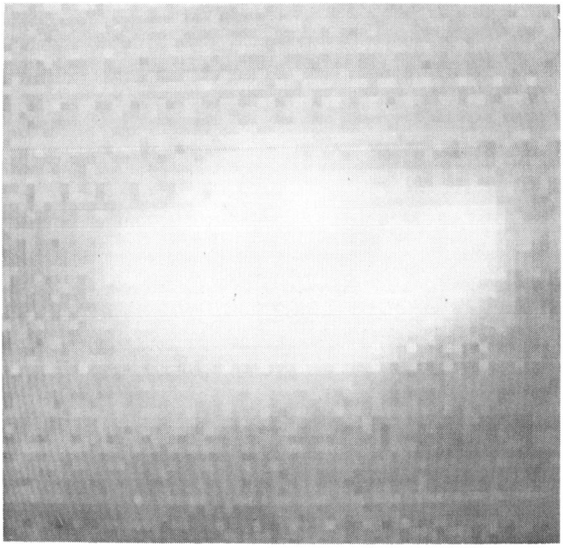

a) Raw data of tank target

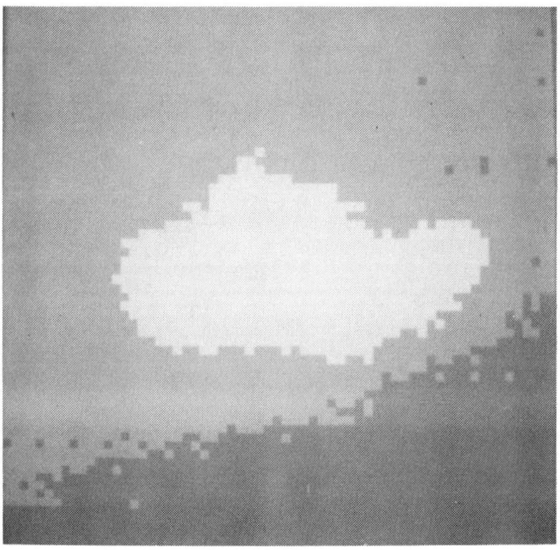

b) Simple rule result (12.517 Sec. used)

Figure 4.24. Detection of tank target 1 experimental results (PDP 11/45 computer was used in the experiment).

c) Context rule result (42.44 Sec. used)

Figure 4.24 (Continued)

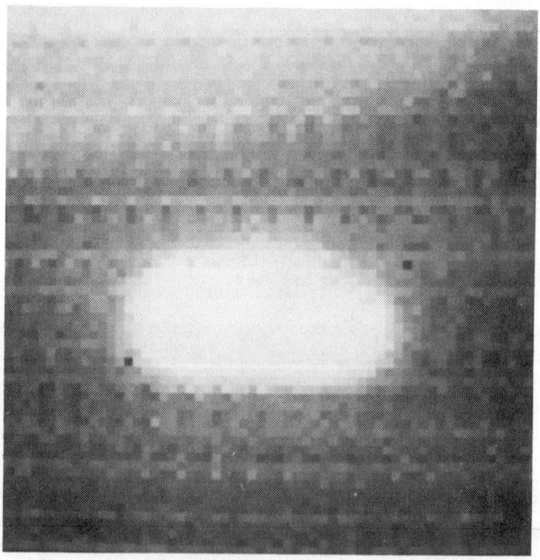

a) Raw data tank target

Figure 4.25. Detection of tank targets experimental results.

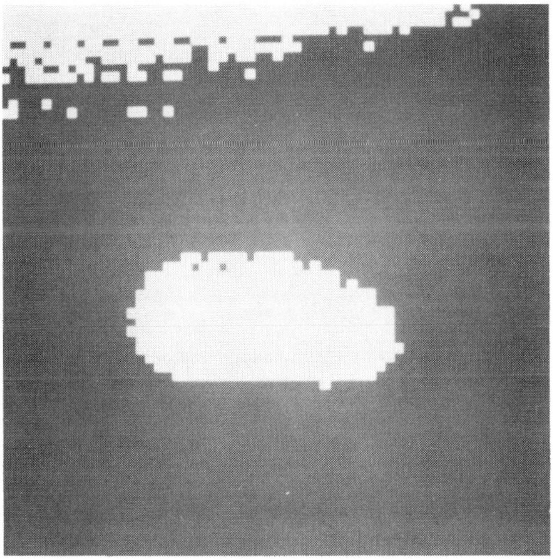

b) Simple rule result

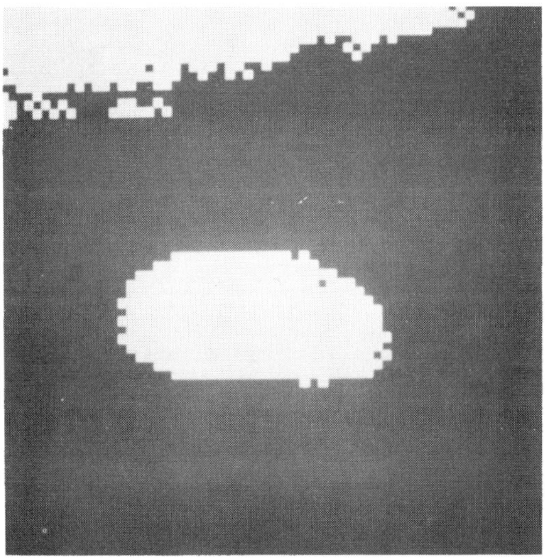

c) Context rule result

Figure 4.25 (Continued)

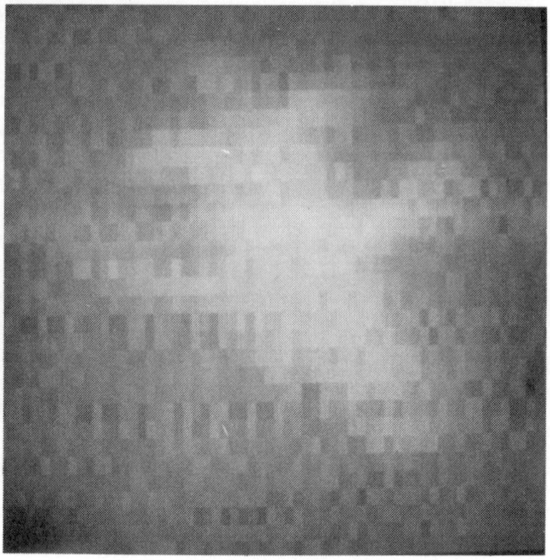

a) Raw data of airplane

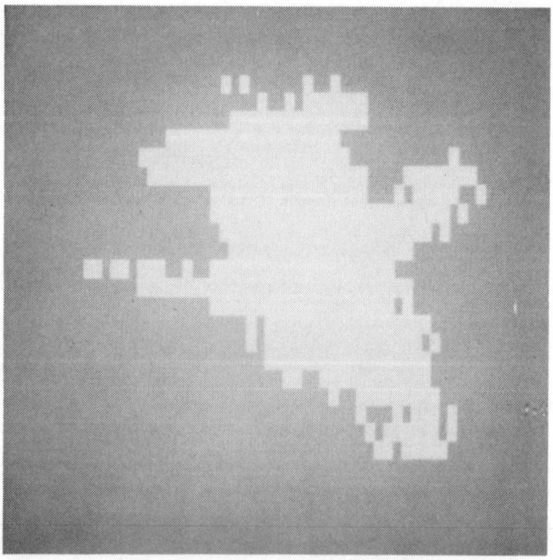

b) Simple rule result

Figure 4.26. Detection of airplane experimental result.

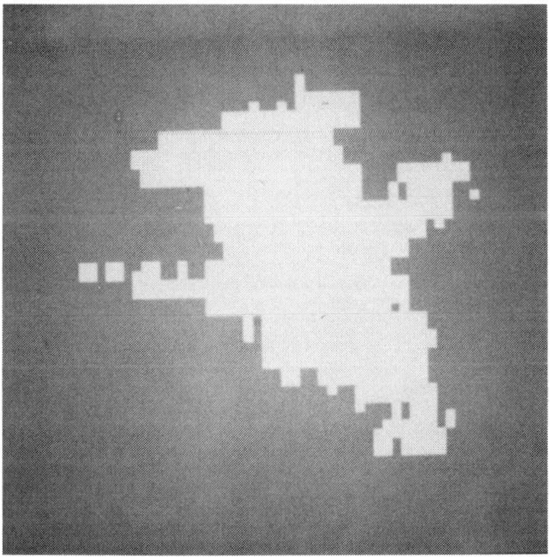

c) Context rule result

Figure 4.26 (Continued)

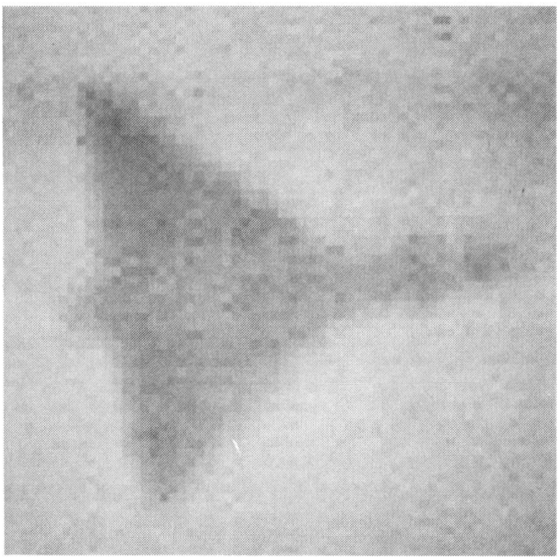

a) Raw data cruise missile

Figure 4.27. Detection of cruise missile experimental results.

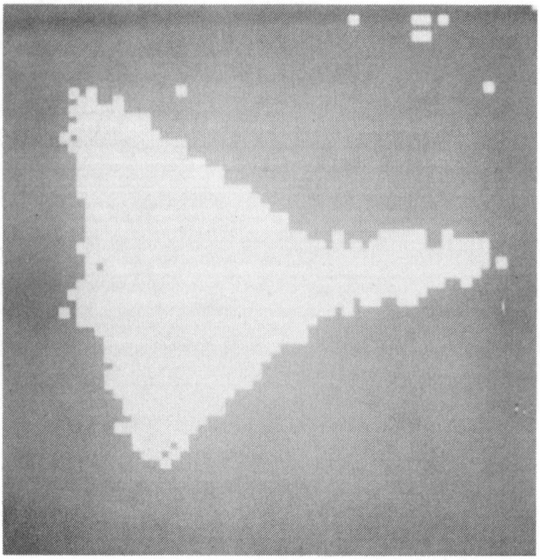

b) Simple rule result

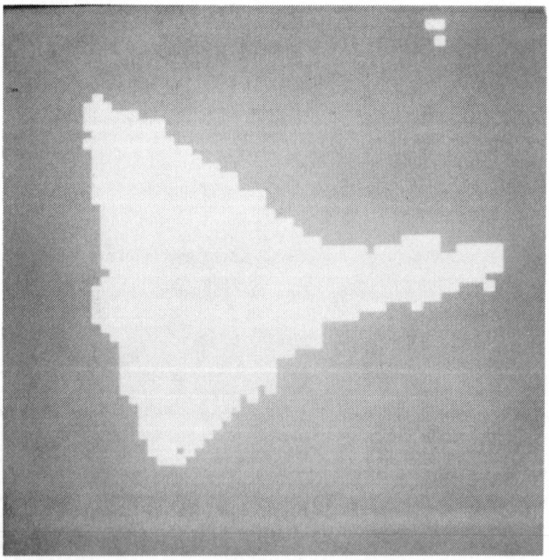

c) Context rule result

Figure 4.27 (Continued)

classifiers are at best awkwardly applied. In addition to the remote sensing applications discussed in previous sections, we would like to demonstrate the capability of the contextual classifier in the detection of tank and airplane targets. The infrared pictorial aircraft data for a tank collected at low attitude is shown in Figure 4.24(a). Because only one band of this data is available, the additional contextual information becomes extremely valuable in the classification process. Training samples are selected within homogeneous background region and target region to obtain mean values and covariances characterizing the target and background. The spatial correlation parameter is estimated from a homogeneous background region and is found to be approximately the same as one used for the remote sensing experiment. The maximum-likelihood classification result is used as the pre-classification result. Only one stage contextual classification is employed since the purpose is to compare the classification performances. The raw data, ML classification results, and one-stage contextual classification results for two tank targets and two cruise missile targets are shown in Figures 4.24, 4.25, 4.26, 4.27 respectively. The contextual classifier again serves effectively to "smooth" the noisy boundaries. Within the homogeneous regions, the contextual classifier also performs better by eliminating the misclassified patterns. Since the computation time (given in the figures) is reasonable, the contextual classifier is a more attractive method than the no-context classifier.

CHAPTER 5
CONCLUSIONS

5.1 Summary of Results

In the present research, most of the efforts have been devoted to develop a feasible stochastic model for contextual pattern classification. The rest of the efforts have been in the implementation of the developed contextual algorithm in the classification of remotely sensed MSS data. The importance of the present research is justified by its theoretical and practical contributions in the field of pattern recognition and remote sensing applications.

First a compound decision scheme is adopted as the preliminary approach to employ contextual information. The true-class directional transitional probability context is used in the compound decision scheme. However, because of the limited ground truth information and the large number of transition probabilities that have to be estimated, the class transition probabilities are not a reliable source of contextual information. The fundamental assumption made in compound decision approach about the independence among all component problems is not appropriate because the class designations of successive samples in MSS remotely sensed data generally are not independent. Also the computational complexity in implementing compound decision rule is so great that this approach can not be used to advantage in practical situations.

Based on the above observations, a new approach to contextual classification is proposed. In order to extract the statistical relationships among the image cells on a two-dimensional image, a stationary stochastic process is defined on the data plane to form a spatial model. To make use of the spatial stochastic model, the remotely sensed data patterns have to be normalized to fit the data to the model. As a result of the normalization process, the contextual information employed come from the correlations among random errors associated with every image cell. Neither the true classes nor the statistics characterizing the true classes contribute any contextual information in the proposed approach. The justification of the existence of correlation among the error variates associated with image cells is given in Section 3.3.4.

In order to normalize pattern vectors, however, the (presumably) true classes of the neighboring cells must be known. Therefore, the requirement of the knowledge about the neighboring true classes is purely for the purpose of fitting our data to the developed model. Since the neighboring classes are usually unknown to us, a preclassification of these neighboring patterns is performed to enable us carrying out the contextual (post) classification. Because of the coding pattern configuration for the NB system, a recursive contextual classification algorithm is proposed when the knowledge about the neighboring classes can be gradually improved and the contextual classification can thus be implemented more effectively.

Experimental results with remotely sensed data demonstrate that substantial improvement in the classification performance can be achieved with the help of contextual information. In particular, the contextual classification results are much more homogeneous within each region. The region boundaries are also well delineated by the contextual classifier showing that the resulting pictures are much smoother in nature.

The computational requirement for the contextual classifier is much less than that of the compound decision rule. In most remote sensing applications, it is believed that two or at most three stages of the recursive contextual algorithm will be enough. This means that the computation time for the contextual classifier is about six to nine times longer than that of the simple decision rule. The computational speed and storage requirements for implementing the contextual classifier appear to be practically feasible.

In the course of present research, we have encountered numerous problems which make the present research far from complete. Therefore, we would like to suggest some new problems for research in this field.

5.2 Suggestions for Further Research

The following research problems which are significant and challenging are suggested. These are also some of the disadvantages of the proposed approach.

a) Since the spatial process defined on the two-dimensional plane is stationary, only the correlations among error variates associated with every image cell are used. It will be a challenging research problem to define a nonstationary spatial stochastic process on the plane such that the different means and covariances characterizing different classes may be used to provide contextual information. To our knowledge, none of the multidimensional nonstationary process has been developed yet. Formulating a decision scheme for this nonstationary model becomes an immediate practical problem.

b) Even though the computation time for the contextual classifier appears to be reasonable, the computational burden will become substantially greater as the number of data points to be classified increases. Therefore, a practical research topic will be to reduce the contextual classification time as much as possible while achieving nearly equivalent performance. The decision tree approach discussed in Chapter 1 aims at reducing the classification time by using multistage classification. When the feature number becomes large, this suggests that perhaps we can incorporate the decision tree approach into the contextual classifier.

c) The extension to the multivariate site-variable situation for the non-stationary model is another theoretical research topic. It is suggested that one must be fairly familiar with the one-dimensional multiple time series problem before taking on such a multidimensional multiple stochastic process.

d) The convergence proof of the recursive classification algorithm does not exist at the time of the present research. It is not known whether the convergence proof will be theoretically possible. However, it will be an interesting and challenging research topic in the future.

REFERENCES

1. Welch, J. R. and K. G. Salter, "A context algorithm for pattern recognition and image interpretation", *IEEE Trans. SMC*, Vol. SMC-1, pp. 24-30, Jan. 1971.
2. Raviv, J., "Decision making in Markov chains applied to the problem of pattern recognition", *IEEE Tran. Inform. Theory*, Vol. IT-13, pp. 536-551, Oct. 1967.
3. Abend, K., "Compound decision procedure for unknown distribution and for dependent state of nature", in *Pattern Recognition*, L. Kanal, Ed., Washington, D.C.: Thompson, pp. 207-249, 1968.
4. Samuel, E., "On simple rules for the compound decision problem", *J. Roy. Stat. Soc.*, Vol. 27, pp. 238-240, 1965.
5. Hussain, A. B. S., "Compound sequential probability ratio test for the classification of statistically dependent patterns," *IEEE Tran. Computers*, Vol. C-23, No. 4, pp. 398-410, April 1974.
6. Chow, C. K., "A recognition method using neighbor dependence", *IRE Tran. on Elec. Comp.*, Vol. EC-11, pp. 683-690, Oct. 1962.
7. Abend, K., "Compound decision procedure for pattern recognition", Proceeding of NEC, Vol. XXII, pp. 777-780, Oct. 1966.
8. Kettig, R. L. and D. A. Landgrebe, "Computer classification of remotely sensed multispectral

image data by extraction and classification of homogeneous objects", LARS Information note 050975, Tech. Report TR-EE 75-24, Purdue University, May 1975.

9. Swain, P. H., "Pattern Recognition: A basis for remote sensing data analysis", LARS Information note 111572, LARS Purdue University, Nov. 1972.
10. Billingsley, P., "Statistical methods in Markov chains", Am. Math. Stat., Vol. 32, pp. 12-40, Mar. 1961.
11. Fu, K. S., Sequential methods in pattern recognition and machine learning, Academic Press, 1968.
12. Fu, K. S., A. Landgrebe, and T. L. Phillips, "Information processing of remotely sensed agricultural data", Proc. IEEE, Vol. 57, pp. 639-653, April 1969.
13. Fu, K. S., "Pattern recognition in Remote Sensing of earth resources", IEEE Tran. on Geo. Scien. and Elec., Jan. 1976.
14. Donaldson, R. W. and T. Toussaint, "Use of contextual constraints in recognition of contour-traced handprinted characters", IEEE Tran. Computers, Nov. 1970.
15. Edwards, A. W. and R. L. Chambers, "Can a priori probabilities help in character recognition?", J. Assoc. Computing Mach., Vol. 11, pp. 465-570, Oct. 1964.
16. Swain, P. H. and R. C. King, "Two effective feature selection criteria for multispectral remote sensing", Proc. Inter. Joint Conf. on Pattern Recognition, Nov. 1973, Washington, D.C.
17. Cary, T. K. and J. C. Lindenlaub, "A case study using LARSYS for analysis of LANDSAT data", LARS Information note 050575, LARS Purdue University.
18. Ferguson, T. S., Mathematical Statistics, Academic Press, Inc., N.Y., 1967.
19. Brook, D., "On the distinction between the conditional probability and the joint probability approaches in the specification of nearest-neighbor systems" Biometrika, 51, 481-483, 1964.

20. Bartlett, M. S., "Physical nearest-neighbor models and non-linear time series," J. Appl. Probl., 8, 222-232, 1971.
21. Bartlett, M. S., "A further note on nearest-neighbor models," J. R. Statist. Soc., A, 131, 579-580, 1968.
22. Bartlett, M. S., "The statistical analysis of spatial patterns," Presented at the Third Conference on Stochastic Processes and their Applications, Sheffield, August 1973.
23. Besag, J. E., "Nearest neighbor system and the auto-logistic model for binary data," J. R. Stat. Soc., B, 34, 75-83, 1972.
24. Besag, J. E., "On the correlation structure of some two dimensional stationary processes," Biometrika, 59, 43-48, 1972.
25. Bartlett, M. S. and Besag, J. E., "Correlation properties of some nearest-neighbor models," Bull. Int. Stat. Inst., 43, Book 2, 191-193, 1969.
26. Grimmett, G. R., "A theorem about random fields," Bull. London Math. Soc., 5, 81-84, 1973.
27. Hammersley, J. M. and Clifford, P., "Markov fields on finite graphs and lattices" (unpublished), 1971.
28. Levy, P., "Chaines double de Markoff et fonctions aleatoires deux variables," C. R. Acad. Sic., Paris, 226, 53-55, 1948.
29. Ord, K., "Estimation methods for Models of Spatial Interaction," J. Amer. Stat. Assoc., Vol. 70, 120-126, 1975.
30. Whittle, P., "Stochastic Processes in Several Dimensions", Bull. Int. Stat. Inst. 40, Book 1, pp. 974-994, 1963.
31. Bartlett, M. S., "Inference and Stochastic Processes," J. R. Stat. Soc., A, 130, 457-577, 1967.
32. Rosanov, Yu A., "On the Gaussian Homogeneous Fields with given conditional distributions," Theor. Prob. Appl., 12, 381-391, 1967.
33. Spitzer, F., "Markov random fields and Gibbs ensembles," Amer. Math. Monthly, 78, 142-154, 1971.

34. Preston, C. J., "Generalized Gibbs states and Markov random fields," Adv. Appl. Probl., 5, 242-261, 1973.
35. Sherman, S., "Markov random fields and Gibbs random fields," Israel J. math., 14, 92-103, 1973.
36. Besag, J. E. and Moran, P. A. P., "On the estimation and testing of Spatial Interaction in Gaussian lattice Processes", Biometrika, pp. 555-562, 1975.
37. Besag, J. E., "Spatial interaction and the statistical analysis of lattice systems," J. R. Stat. Soc., B 36, 192-236, 1974.
38. Bartlett, M. S., "Physical nearest-neighbor models and non-linear time series, II Further discussion of approximate solutions and exact equations," J. Appl. Probl, 9, 76-86.
39. Besag, J. E., "Nearest-neighbor systems: A lemma with Applications to Bartlett's Global Solutions," J. Appl. Prob., 9, 418-421, 1972.
40. Moran, P. A. P., "A Gaussian Markovian Processes on a lattice," J. Appl. Prob., 10, 605-612, 1973.
41. Moran, P. A. P., "Necessary conditions for Markovian Processes on a lattice," J. Appl. Prob., 10, 605-612, 1973.
42. Whittle, P., "On stationary processes in the plane," Biometrika, 41, 434-449, 1954.
43. Wacker, A. G. and Landgrebe, D. A., "The minimum distance approach to classification," LARS information note 100771, LARS, Purdue University, 1971.
44. Anderson, T. W., An introduction to Multivariate Statistical Analysis, Wiley, 1958.
45. Rao, C. R., Linear Statistical Inference and its Application, Wiley, 1965.
46. Roy, S. N., Some Aspects of Multivariate Analysis, Wiley, 1957.
47. Kshirsagar, A. M., Multivariate Analysis, Marcel Dekker, 1972.
48. Johnson, N. L. and S. Kotz, Distributions in Statistics, Continuous Multivariate Distributions, Wiley, 1972.
49. Abramowitz, M. and Stegun, I. A., Handbook of

Mathematical Functions, Dover, New York, 1965.
50. Katsura, S. and Inawashiro, S., "Lattice Green's function for the rectangular and square lattices at arbitrary points," J. Math. and Phys. 12, 1622-1630, 1971.
51. Levy, P., Processus Stochastiques et Movement Brownien, Gautier Villars, Paris, 1948.
52. Kolmogorov, A., Foundations of probability theory, New York, 1956.
53. Clifford, P. and A. Sudbury, "A model for spatial conflict," Biometrika, 60, 3, pp. 581-588, 1973.
54. Hammersley, J. M., "Stochastic models for the Distribution of particles in Space," Suppl. Adv. Appl. Prob., pp. 47-68, 1972.
55. Bartlett, M. S., "Physical nearest-neighbor models and non-linear time series: III Non-zero means and sub-critical temperatures," J. Appl. Prob., 11, pp. 715-725, 1974.
56. Averintsev, M. B., "On a method of describing complete parameter fields," Problemy peredaci Informatsii, 6, 100-109, 1970.
57. Montroll, E. W., "Random walks on lattices," Proceedings of Symposia in Applied Mathematics, 16, 193-200, 1964.
58. Bauer, M. E. and J. E. Cipra, "Identification of agricultural crops by computer processing of ERTS MSS data," Proceedings of the symposium on significant results obtained from ERTS-1, Goddard Space Flight Center, Greenbelt, Maryland, March 5-9, 1973, Vol. 1, Section A, pp. 205-212.
59. Mausel, Todd, M. F. Baumgardner, "Analysis of Metropolitan Land-use by machine processing of Earth Resources Technology Satellite data", LARS information note 031276.
60. Todd, W. J., M. F. Baumgardner, "Land-use classification of Marion County, Indiana by spectral analysis of digitized satellite data", LARS information note 101673.
61. Swain, P. H., "Land-use classification and mapping by machine-assisted analysis of LANDSAT MSS data", Final report on U.S. Geological Survey Contract No. 14-08-0001-14725, LARS Infor-

mation Note 111276, LARS, Purdue University, W. Lafayette, IN.

62. Anderson, J., E. Hardy and J. Roach, "A land-use classification system for use with remote sensor data", U.S. Geological Survey Circular 671, (Revised version) Washington, D.C., 1976, (Geological survey professional paper 964).

63. Kettig, R. L., and D. Landgrebe, "Classification of Multispectral image data by extraction and classification of homogeneous objects," _IEEE Trans. Geo. Sci. Elect._, Vol. GE-14, Jan. 1976, pp. 19-26.

64. Anuta, P., M. Bauer, "An analysis of temporal data for crop species classification and Urban change detection", LARS Information Note 110873.

65. Kawamura, J. G., "Automatic recognition of changes in Urban development from Aerial photographs", _IEEE Trans. on SMC_, Vol. SMC-1, No. 3, pp. 230-239, July 1971.

66. Ellefsen, R., P. H. Swan, James Wray, "Urban Land-use Mapping by machine processing of ERTS-1 MSS data: A San Francisco Bay Area example", LARS Information Note 101573.

67. Robertson, T. V., K. S. Fu, P. H. Swain, "Multispectral Image Partitioning", Tech. Rept. TR-EE 73-26, School of Electrical Engineering, Purdue University, W. Lafayette, Ind., August 1973.

68. Anuta, P. E., "Spatial Registration of Multispectral and Multitemporal Imagery Using Fast Fourier Transform Techniques", _IEEE Trans. Geoscience Electronics_, Vol. GE-8, No. 4, pp. 353-368, Oct. 1970.

69. Rodd, E. M., "Closed Boundary Field Selection in Multispectral Digital Images", IBM Publications No. 320, 2420, Jan. 1972.

70. Muerle, J. L. and D. C. Allen, "Experimental Evaluation of Techniques for Automatic Segmentation of Objects in a Complex Scene", _Pictorial Pattern Recognition_, G. C. Cheng, R. S. Ledley, D. K. Pollock, A. Rosenfeld (eds.), Thompson Book Co., Washington, D.C., pp. 3-13, 1968.

71. Robertson, T. V., "Extraction and classification of Objects in Multispectral Images", Proceedings of the Conference on Machine Processing of Remotely Sensed Data, Purdue University, W. Lafayette, Ind., Section 3B, pp. 27-34, Oct. 1973.

72. Kettig, R. L. and D. A. Landgrebe, "Automatic Boundary Finding and Sample Classification of Remotely Sensed Multispectral Data", LARS Information Note 041773, Purdue University, W. Lafayette, Ind., April 1973.

73. Gupta, J. N. and P. A. Wintz, "Closed Boundary Finding, Feature Selection, and Classification Approach to Multi-Image Modeling", LARS Information Note 062773, Purdue University, W. Lafayette, Ind., June 1973.

74. Mendel, J. M. and K. S. Fu, eds., <u>Adaptive, Learning and Patter Recognition Systems: Theory and Applications</u>, Academic Press, New York, 1970.

75. Duda, R. O. and P. E. Hart, <u>Pattern Classification and Scene Analysis</u>, Wiley, New York, 1973.

76. Chen, C. H., <u>Statistical Pattern Recognition</u>, Hayden, Washington, D. C., 1973.

77. Nagy, G., "Digital Image-Processing Activities in Remote Sensing for Earth Resource", <u>Proc. of IEEE</u>, Vol. 60, pp. 1177-1200, Oct. 1972.

78. Fu, K. S. and A. Rosenfeld, "Pattern Recognition and Image Processing," <u>IEEE Trans. Computers</u>, Vol. C-25, 1336-1345, Dec. 1976.

79. Ho, Y. C. and A. K. Agrawala, "On Pattern Classification Algorithms Introduction and Survey", <u>Proc. of IEEE</u>, Vol. 56, pp. 2101-2114, Dec. 1968.

80. Kanal, L. N., "Interactive Pattern Analysis and Classification Systems: A Survey and Commentary", <u>Proc. of IEEE</u>, Vol. 60, pp. 1200-1215, Oct. 1972.

81. Meisel, W. S., <u>Computer-Oriented Approaches to Pattern Recognition</u>, Academic press, 1972.

82. Fukunaga, K., <u>Introduction to Statistical Pattern Recognition</u>, Academic Press, 1972.

83. Holmes, R. A. and R. B. MacDonald, "The Physi-

cal basis of system design for remote sensing in agriculture", Proc. of IEEE, Vol. 57, pp. 629-639, 1969.

84. Fleming, M., J. Berkebile, and R. Hoffer, "Computer-aided Analysis of LANDSAT-1 MSS Data: A comparison of three approaches including the 'modified clustering' Approach", Proc. of the symposium on Machine Processings of Remotely Sensed Data, Purdue University, W. Lafayette, Ind., June 3-5, 1975, pp. 1B-54 to 1B-61.

85. Hoffer, R. M. and Staff, "Computer-aided analysis of SKYLAB MSS data in Mountainous Terrain for Land Use, Forestry, Water Resource, and Geologic applications", SKYLAB Final Report, LARS Information Note 121275, LARS Purdue University, W. Lafayette, Ind., pp. 381.

86. Wu, C., D. Landgrebe, and P. Swain, "The decision tree approach to classification", School of Electrical Engineering, Tech. Rept. TR-EE 75-17, Purdue University, W. Lafayette, Ind., May 1975.

87. Nilsson, N. J., Problem Solving Methods in Artificial Intelligence, McGraw-Hill, 1971.

88. You, K. C. and Fu, K. S., "An approach to the design of a linear binary tree classifier", Proc. of the Symposium on Machine Processings of Remotely Sensed Data, Purdue University, W. Lafayette, Ind., June 29-July 1, 1976, pp. 3A-1 to 3A-10.

89. Robbins, H., "Asymptotically subminimax solutions of compound statistical decision problems," Proc. Second Berkeley Symp. Math. Statist. Prob., 131-148, 1958.

90. Samuel, E., "On simple rules for the compound decision problem," J. Roy. Stat. Soc., B, 27, 238-244, 1965b.

91. Robbins, H., "Some numerical results on a compound decision problem," Recent Development in Information and Decision Processes, Edited by R. Machol and P. Gray, MacMillan Company, 56-62, 1962.

92. Robbins, H., "The empirical Bayes approach to statistical decision problems," Ann. Math. Stat., 35, 1-20, 1964.

93. Samuel, E., "Sequential compound rules for the finite decision problem," J. Roy. Stat. Soc., B, 28, 63-72, 1966.
94. Samuel, E., "Sequential compound estimators," Ann. Math. Stat., 36, 879-889, 1965a.
95. Samuel, E., "Asymptotic solutions of the sequential compound decision problem," Ann. Math. Stat., 34, 1079-1091, 1963a.
96. Samuel, E., "Convergence of the losses of certain decision rules for the sequential compound decision problem," Ann. Math. Stat., 35, 1606-1621, 1964.
97. Van Ryzin, J. R., "The sequential compound decision problem with m x n finite loss matrix," Ann. Math. Stat., 37, 954-975, 1966b.
98. Johns, M. V., Jr., "Two-action compound decision problems," Technical Report No. 87, Stanford University, 1966.
99. Neyman, J., "Two breakthroughs in the theory of statistical decision making," Rev. Inst. Int. Stat., 30, 11-27, 1962.
100. Robbins, H., "An empirical Bayes approach to statistics," Proc. Third Berkeley Symp. Math. Statist. Prob., 1, 157-164, 1955.
101. Johns, M. V. Jr., "Non-parametric empirical Bayes procedures," Ann. Math. Stat., 28, 649-669, 1957.
102. Johns, M. V. Jr., "An empirical Bayes approach to non-parametric two-way classification," Studies in Item Analysis and Prediction, H. Salomon, editor, Stanford Universty Press, 221-232, 1961.
103. Miyasawa, K., "An empirical Bayes estimator of the mean of a normal population," Bull. Inst. Int. Stat., 38, 181-188, 1961.
104. Samuel, E., "An empirical Bayes approach to the testing of certain parametric hypotheses," Ann. Math. Stat., 34, 1370-1385, 1963b.
105. Samuel, E., "The compound decision problem in the opponent case," Israel J. Math, 3, 117-126, 1965c.
106. Hanan, J. F. and Van Ryzin, J. R., "Rate of convergence in the compound decision problem for two completely specified distributions,"

Ann. Math. Stat., 36, 1743-1752, 1965.

107. Hannan, J. F. and Robbins, H., "Asymptotic solutions of the compound decision problem for two completely specified distributions," _Ann. Math. Stat._, 26, 37-51, 1955.

108. Van Ryzin, J. R., "The compound decision problem with m x n finite loss matrix," _Ann. Math. Stat._, 37, 412-424, 1966a.

109. Swain, D. D., "Bounds and rates of convergence for the extended compound estimation problem in the sequence case," Technical Report No. 81, Stanford University, 1965.

110. Barndorff-Nielsen, O., "On the empirical Bayes methodology for stochastically dependent parameters," Unpublished, 1964.

111. Noble, B., _Applied Linear Algebra_, Prentice Hall, 1969.

112. Swain, P. H. and H. Hauska, "The decision tree classifier: Design and potential", _IEEE Tran. on Geos. Elec._, Vol. GE-15, No. 3, July 1977, pp. 142-146.

113. Kulkarni, A. V. and L. N. Kanal, "An Optimization Approach to Hierarchical Classifier Design," Proc. Third Int'l. Conf. Pattern Recognition, Dec. 1976, Coronado, Calif.

114. Doob, J. L., _Stochastic Processes_, Wiley,

APPENDIX A
Derivation of a Sequential Compound Decision Rule

If only the first k observations, $\underline{X}_k = (x_1,\ldots,x_k)$ are available when the kth decision must be made, one can use a sequential compound decision rule $\underline{t}_N = (t_1,\ldots,t_N)$, where $t_k = t_k(j/\underline{X}_k)$ is for each $\underline{X}_k$ a distribution over A according to which the kth action is chosen. The risk for such a rule is given by Equation (2.14) of Section 2.1.3. The kth component risk is

$$r(\underline{\theta}_N,t_k) = \int \sum_{j=1}^{S} L_{\theta_k} t_k(j/X_k)p(X_k/\underline{\theta}_k)dX^k = r(\underline{\theta}_k,t_k) \quad (A1)$$

We assume that $p_i(x)$ is known for all $i \in \Omega$, but that none of the θ_k's is known. The compound Bayes risk with respect to an a priori distribution $G(\underline{\theta}_N)$ over Ω^N is given by equations (2.7) and (2.8). A procedure is compound Bayes against G when it minimizes (2.7). Thus, the sequential compound Bayes procedure $\underline{t}_N^G$ is the one that, for every k, minimizes the kth component Bayes risk,

$$r(G,t_k) = \sum_{\theta_N} r(\underline{\theta}_k,t_k)G(\underline{\theta}_N) = \sum_{\theta_k} r(\underline{\theta}_k,t_k)G(\underline{\theta}_k)$$

$$= \sum_{i=1}^{s} \sum_{\underline{\theta}_k} L_{\underline{\theta}_k} j p(X_k/\underline{\theta}_k)G(\underline{\theta}_k)t_k(j/X_k)dX^k \quad \text{(A2)}$$

where $G(\underline{\theta}_k)$ is the (marginal) a priori distribution over Ω^k. Hence $t_k^G(j/X_k) = 1$ for that j which minimizes the quantity

$$Q = \sum_{\underline{\theta}_k} L_{\theta_k} j p(\underline{X}_k/\underline{\theta}_k)G(\underline{\theta}_k) \quad \text{(A3)}$$

Theoretically, the problem is solved. Practically, it is not. We have a sum of r^k terms, where k may be in the thousands. Let $p(\underline{X}_k,\underline{\theta}_k) = p(\underline{X}_k/\underline{\theta}_k)G(\underline{\theta}_k)$, then

$$Q = \sum_{\theta_k} L_{\theta_k} j p(\underline{X}_k,\underline{\theta}_k) \quad \text{(A4)}$$

We note that for the special case where action j corresponds to deciding that $\theta = j$ and

$L_{\theta_J} = \begin{cases} 1 \text{ if } \theta \neq j \\ 0 \text{ if } \theta = j \end{cases}$ t_k^G chooses the value of θ_k that maximizes $p(\underline{X}_k,\theta_k)$. This is equivalent to maximizing the a posteriori probability $G(\theta_k/\underline{X}_k)$.

In Equation (A4)

$$p(\underline{X}_k,\theta_k) = p(x_k/\theta_k)p(\underline{X}_{k-1},\theta_k) = p(x_k/\theta_k) \sum_{\underline{\theta}_{k-1}} p(\underline{X}_{k-1},$$

$$= p(x_k/\theta_k) \sum_{\underline{\theta}_{k-1}} p(\underline{X}_{k-1}/\underline{\theta}_{k-1})G(\underline{\theta}_k)$$

$$= p(x_k/\theta_k) \sum_{\underline{\theta}_{k-1}} G(\theta_k/\underline{\theta}_{k-1})p(\underline{X}_{k-1},\underline{\theta}_{k-1}) \quad \text{(A5)}$$

If the states of nature form a Markov chain,

$$G(\theta_k/\underline{\theta}_{k-1}) = G(\theta_k/\theta_{k-1})$$

with known transition probabilities $G(w/v)$, for w and v in Ω, then

$$p(\underline{X}_k,\underline{\theta}_k) = p_{\theta_k}(X_k) \sum_{\theta_{k-1}=1}^{r} G(\theta_k/\theta_{k-1})p(\underline{X}_{k-1},\theta_{k-1}) \quad (A7)$$

Note that $p_i(x)$ and $G(i/v)$ are known and $p(\underline{X}_{k-1},\theta_{k-1})$ is defined recursively in terms of $p(x_1,\theta_1) = p_{\theta_1}(x_1)G(\theta_1)$. The kth decision $t_k^G(j/\underline{X}_k)$, depends on x_k and a function of θ_{k-1} which had already been calculated in order to make the previous decision. It does not depend on the (unknown) value of θ_{k-1} nor on the decision about it. This decision rule was developed independently by Abend (3) and Raviv (2).

APPENDIX B
An 8-Neighbor Decision Rule Derivation

In the following, an 8-neighbor compound decision rule derivation is presented for employing contextual information in improving classification accuracy. As mentioned before, the expression $p(\underline{X}_n/\theta_k)$ is a high-dimensional vector and thus making evaluation of this density function is impossible without making some simplifying assumptions.

k_7	k_2	k_6
k_3	k	k_1
k_8	k_4	k_5

Figure B1. Labeling of Cells of 8-Neighbor System.

For every cell k, define adjacent cells as any two cells with a common side. The first assumption is that contextual relationships between non-adjacent cells are negligible. Thus for any non-adjacent cells b and c,

$$p(x_b/\theta_c,\text{any other } x \text{ or } \theta)=p(x_b/\text{any other } x \text{ or } \theta) \quad \text{(B1)}$$

The second assumption is that the appearance x_k of a class θ_k is a function only of θ_k, and if θ_k is known, neither the nature nor the appearance of any other pixel provides additional information about x_k. This can be expressed as

$$p(x_k/\theta_k, \text{any other } \theta \text{ or } x) = p(x_k/\theta_k) \qquad (B2)$$

The above two assumptions imply the following

$$p(x_b/x_c, \text{any other } x \text{ or } \theta) = p(x_b/\text{any other } x \text{ or } \theta) \qquad (B3)$$

if b and c are non-adjacent cells.

The third assumption is that the contribution due to corner cells k_5, k_6, k_7, k_8 is independent of that due to the four adjacent cells to k. This is a reasonable assumption since the four corner cells are of a distance $\sqrt{2}$ times longer to cell k than from four adjacent cells (see Figure B1).

Referring to Figure B1 we obtain

$$p(\underline{x}_n/\theta_k) = p(x_{NA}, x_{AD}, x_{CO}, x_k/\theta_k)$$

$$= p(x_{NA}/x_{AD}, x_{CO}, x_k, \theta_k)p(x_{AD}, x_{CO}, x_k/\theta_k) \qquad (B4)$$

where $\underline{x}_n$ = all the measurement vectors in the frame

x_k = vector of cell k

x_{AD} = $(x_{k_1}, x_{k_2}, x_{k_3}, x_{k_4})$, a collection of vectors from the four cells adjacent to k.

x_{NA} = $(x_{k_9}, x_{k_{10}}, \ldots, x_{k_{128^2}})$, all non-adjacent cells to k

x_{CO} = $(x_{k_5}, x_{k_6}, x_{k_7}, x_{k_8})$, a collection of vectors from the four corner cells of k

Since the cells of x_{NA} are not adjacent to cell k, we use (B1) and (B3) to obtain

$$p(\underline{x}_N/\theta_k) = p(x_{NA}/x_{AD}, x_{CO})p(x_{AD}, x_{CO}, x_k/\theta_k) \qquad (B5)$$

The optimum decision rule can be restated as: choose $a \in A$ to minimize

$$\sum_{\text{all } \theta_k} L(\theta_k, a) p(x_{NA}/x_{AD}, x_{CO}) p(x_{AD}, x_{CO}, x_k/\theta_k) G(\theta_k) \quad (B6)$$

Since $p(x_{NA}/x_{AD}, x_{CO})$ does not involve cell k and so is not a function of θ_k, it can be put outside the summation; and since it is not a function of a, it can not affect the decision.

The new rule can be restated as: choose $a \in A$ to minimize

$$\sum_{\text{all } \theta_k} L(\theta_k, a) p(x_{AD}, x_{CO}, x_k/\theta_k) G(\theta_k)$$

Further simplifications are still needed for the evaluation of a joint density function for eleven such vectors.

First we use the third assumption as follows:

$$p(x_{AD}, x_{CO}, x_k/\theta_k) = p(x_{AD}, x_k/\theta_k) p(x_{CO}, x_k/\theta_k)$$

$$= p(x_k/x_{AD}, \theta_k) p(x_{AD}/\theta_k) p(x_k/x_{CO}, \theta_k) p(x_{CO}/\theta_k) \quad (B7)$$

But $p(x_{AD}/\theta_k) = p(x_{k_1}, x_{k_2}, x_{k_3}, x_{k_4}/\theta_k)$

$$= p(x_{k_1}/x_{k_2}, x_{k_3}, x_{k_4}, \theta_k) p(x_{k_2}/x_{k_3}, x_{k_4}, \theta_k)$$

$$p(x_{k_3}/x_{k_4}, \theta_k) p(x_{k_4}/\theta_k)$$

$$= p(x_{k_1}/\theta_k) p(x_{k_2}/\theta_k) p(x_{k_3}/\theta_k) p(x_{k_4}/\theta_k)$$

$$= \prod_{i=1}^{4} p(x_{k_i}/\theta_k)$$

Next expanding each $p(x_i/\theta_k)$ in terms of θ_i and using (B2), we obtain

$$p(x_i/\theta_k) = \sum_{\theta_i} p(x_i,\theta_i/\theta_k)$$

$$= \sum_{\theta_i} p(x_i/\theta_i,\theta_k)p(\theta_i/\theta_k)$$

$$= \sum_{\theta_i} p(x_i/\theta_i)p(\theta_i/\theta_k)$$

Then

$$p(x_{AD},x_k/\theta_k)=p(x_k/\theta_k)\prod_{i=1}^{4}\sum_{\theta_{k_i}} p(x_{k_i}/\theta_{k_i})p(\theta_{k_i}/\theta_k) \quad (B8)$$

Similarly, $p(x_{CO},x_k/\theta_k) = p(x_k/\theta_k)p(x_{CO}/\theta_k)$

$$p(x_{CO}/\theta_k) = p(x_{k_5},\ldots,x_{k_8}/\theta_k)$$

$$= \prod_{i=5}^{8} p(x_{k_i}/\theta_k)$$

$$p(x_{k_i}/\theta_k) = \Sigma_{\theta_i} p(x_{k_i}/\theta_i)p(\theta_i/\theta_k)$$

Then

$$p(x_{CO},x_k/\theta_k)=p(x_k/\theta_k)\prod_{i=5}^{8}\sum_{\theta_{k_i}} p(x_{k_i}/\theta_{k_i})p(\theta_{k_i}/\theta_k) \quad (B9)$$

The decision rule now is to choose $a \in A$ to minimize

$$\sum_{\text{all } \theta_k} L(\theta_k,a)p(x_k/\theta_k) \underbrace{\prod_{i=1}^{4}\sum_{\theta_{k_i}} p(x_{k_i}/\theta_{k_i})p(\theta_{k_i}/\theta_k)}_{\text{adj}}$$

$$\cdot\ p(x_k/\theta_k) \underbrace{\prod_{i=5}^{8} \sum_{\theta_{k_i}} p(x_{k_i}/\theta_{k_i})\ p(\theta_{k_i}/\theta_k)}_{\text{con}} \cdot G(\theta_k) \qquad \text{(B10)}$$

This is the contextual rule in its final form. Note that "adj" term denotes the contribution due to four adjacent elements whereas "con" term denotes the contribution due to four corner points of cell k.

APPENDIX C
An Example of Using Newton-Raphson Technique

We would like to find $\hat{p}$ which minimizes

$$\left(\prod_{i=1}^{n}(1 - p\lambda_i)\right)^{-2/n}(y'y - 2py'y_L + p^2 y_L'y_L) \qquad (C1)$$

where y, y_L are n x 1 vectors, p, λ are scalar.

The ML estimator is that value of p which minimizes

$$f(p) = -\frac{2}{n}\sum_{i=1}^{n} \ln(1 - p\lambda_i) + \ln(s^2) \qquad (C2)$$

where

$$s^2 = s^2(p) = y'y - 2py'y_L + p^2 y_L'y_L$$

and f(p) is the logarithm of expression (C1). From (C2), the derivatives of f(p) are

$$f_p(p) = \frac{2}{n}\sum_{i=1}^{n} \lambda_i/(1-p\lambda_i) + 2(py_L'y_L - y'y_L)/s^2 \qquad (C3)$$

and

$$f_{pp}(p) = \frac{2}{n} \sum_{i=1}^{n} (\lambda_i)^2/(1 - p\lambda_i)^2 + 2y_L{}'y_L/s^2$$

$$- 4(py_L{}'y_L - y'y_L)^2/s^4 \quad .$$

Then $\hat{p}$ may be determined iteratively from the usual expression

$$p_{(r+1)} = p_{(r)} - f_p(p_{(r)})/f_{pp}(p_{(r)}) \qquad \text{(C5)}$$

A useful starting value is $p_{(0)} = y'y_L/y'y$.

Appendix D
Computer Flow Charts

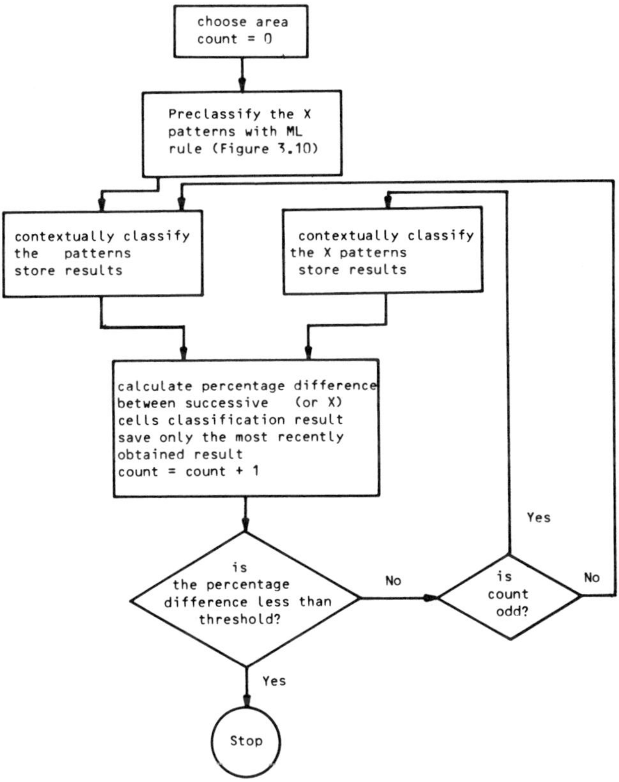

Figure D1. Block diagram of single-stage contextual classification experiment.

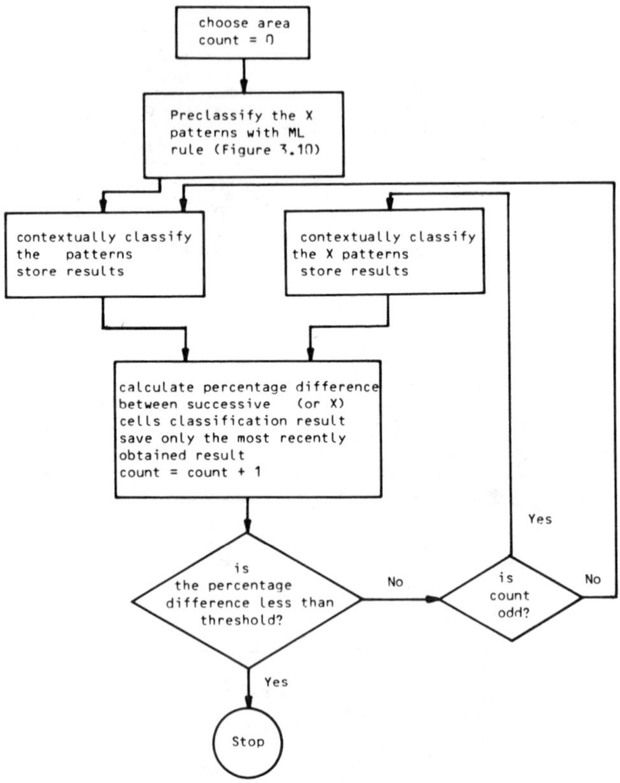

Figure D2. Flow chart of recursive contextual classification algorithm.

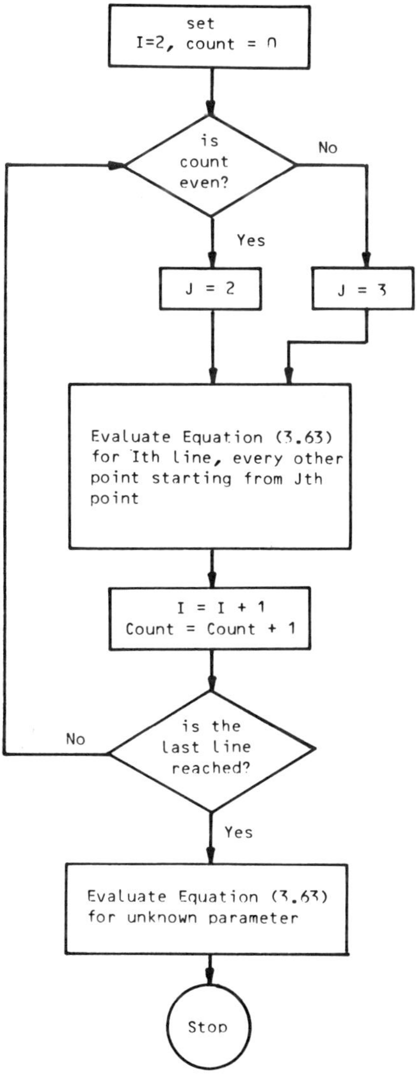

Figure D3. Flow chart of coding techniques for parameter estimation.